The voyage of the *Emigrant*, 1844–1847. Traced on *Colton's Map of the World*, 1855. (Courtesy Wheaton College Library)

Prague
Olmutz
Breslau
Cracow
Lublin
Pilsen
Vienna
Munich
Austria
Buda
Trent
Gross Wardein
Arad
Temesvar
Peterwarden
Trieste
Bukarest
Mostar
Nikopol
Silistria
Adrianople
Salonica
Orleans
Tours
Paris
Havre
Calais
Berne
Lyons
Geneva
Parma
Genoa
Toulouse
Bayonne
Marseille
Corsica
Barcelona
Bordeaux
Bay of Biscay
Santander
Bilbao
Santiago
Spain
Madrid
Ciudad Real
Seville
Lagos
Cadiz
Gibraltar
Ceuta
Fez
Saffie
Marocco
Tafile
Sea of Azof
Astrachan
Kresta
Crimea
Sevastopol
Georghievsk
Black Sea
Constantinople
Eregli
Sibas
Trebizond
Erzeroum
Akka
Nun
Agabli
Oasis of Twat
Ghadamis
Ghat
Murzuk
Fezzan
Zella
Sywah
Egypt
Wah el Dakhel
Febabo
Ru. of Thebes
El Der
Petrea
Ouejeh
Aneyzeh
El Katif
Medina
Derayeh
Fouf
Yambo
Mecca
Tayf
Arabia
Komol
Selimeh
Nubia
New Dongola
Meroe
Berber
Kartoum
Koomfidah
Loheia
Mocha
Aden
Str. of Babel Mandeb
Rostak
Muscat
Ainad
Dofar
Shiraz
Great Desert of Sahara
Towaber
Tisheet
Timbuctoo
Bilma
Air
Agadez
Niger
Benown
Kemmoo
Jenne
Coro
Sakatu
Katshna
L. Tsad
L. Fittri
Cobbe
Wara
Axum
Gondar
Sennar
Abyssinia
Blue R.
Tajoura
Kobbo
Adel
Ajan
Sego
Timbo
Boussa
Kuka
Tubelda
Soudan
Banda
Abomey
Benin
N. Guinea
Coomassie
Liberia
Ecrikok
Dar Kulla
Cameroons Mt. & R.
Biafra
Bight of Benin
Fernando Po
Bight of Biafra
Cape Coast C.
Shiwella
Bedouin
Bassas
Galweno
Mukdeesha
Juba
Patta
Melinda
Mombas
Zanzibar I. & T.
Zanzibar
Quiloa
Enengo
Cassabella
Loango
Unexplored
Region
L. Nyassi
Congo
Ambriz
Angola
Loanda
Cazembe
R. Livuma
Catende
Benguela
Caconda
Dumbo
Mozambique
Mozambique Channel
Madagascar
Senna
Sofala
St. Helena
Cimbeba
L. Ngami
Bechuanas
Damara
Sandy coast destitute of water
Inhambane
Delagoa B.
Santa Cruz
Hottentots
Melita
Africaners Krall
Orange R.
Lily
Cape Colony
Tulbagh
Cape Town
Cape of Good Hope
Bloemfontein
Natal
Pietermaritzburg
Port Natal
Ascension I.
Atlantic Ocean
Indian Ocean

THE AMERICAN MARITIME LIBRARY: VOLUME III

"There She Blows"

"On the look out." Color lithograph by Currier & Ives. (Courtesy Kendall Whaling Museum)

"There She Blows:"

A

NARRATIVE

OF A

WHALING VOYAGE,

IN THE INDIAN AND SOUTH ATLANTIC OCEANS

by

BEN-EZRA STILES ELY

Edited, with an introduction, notes, and appendices,

by

CURTIS DAHL

Published for

THE MARINE HISTORICAL ASSOCIATION, INCORPORATED

by

WESLEYAN UNIVERSITY PRESS

Middletown, Connecticut

ISBN: 0-8195-4032-3
Library of Congress Catalog Card Number: 76-142726
Manufactured in the United States of America
First edition

In Memory of
Laura Ely Curtis
and
Elizabeth Curtis Dahl

Contents

List of Illustrations

Foreword

THIS edition of *"There She Blows:" A Narrative of a Whaling Voyage, in the Indian and South Atlantic Oceans*, by Ben Ezra Stiles Ely, is a reprint of a little book of 107 pages published originally in Philadelphia in 1849. To the original text has been added a biographical introduction, a chapter from Ely's manuscript autobiography incorporating the account of his religious conversion at sea, excerpts from two fortunately preserved manuscript journals detailing the events of the same voyage on the same ship, and detailed information on the vessel, her course, and her officers and crew. Thus this volume, in addition to reprinting a piquant book of whaling experience, brings together a great deal of the recoverable information about a representative whaling voyage. It presents both the facts and Ely's rendering of them.

The book *"There She Blows"* is itself a rare one. Though there may well be others, I have after considerable search located only seven copies. Three of these are in university libraries: at Princeton, Columbia, and Brown. One is in the library of the American Antiquarian Society in Worcester. A fifth is in the collection of Charles Batchelder, the well-known scholar of whaling lore. A sixth is the copyright copy in the Library of Congress. The last is my own, which I inherited through my mother from her mother, a daughter of the author. Another copy was supposedly offered for sale about 1960 by a dealer in Windsor, Connecticut. Its whereabouts, if it exists, I have been unable to ascertain.

Physically the book is a small one, measuring only six inches high by four inches wide. It was printed by William F. Geddes, 112 Chestnut Street, Philadelphia, and published by James K. Simon, a bookseller doing business at the southeast corner of Fifth and Spruce Streets. The copy in my possession is bound in spotted green. On the front cover is pasted a black label with a delightful picture of a spouting whale stamped on it in gold—a picture that has been reproduced for this edition.

My thanks are particularly due to those who have permitted me to reprint manuscript materials in their possession. I am indebted to Ben Ely (Ben Ezra Stiles Ely, III) of Hannibal, Missouri, for his courteous permission to reprint parts of his grandfather's manuscript autobiography now in his possession; to the Rhode Island Historical Society for permission to print parts of the unpublished journal of Seth F. Lincoln and also various documents from the Bristol Customs House now in their possession; and to the Marine Museum of Fall River for permission to print excerpts from the journal of Charles F. Tucker, on loan to the Museum from Mrs. Sumner J. Waring, Sr., of Fall River.

Foreword

For generous help in preparing this volume I am indebted to Reginald Hegarty, curator of the Melville Room of the New Bedford Free Public Library; Charles F. Batchelder, who gave me the clues by which to track down the Tucker Journal and has helped with advice; David H. Atwater, Jr., of the Marine Museum at Fall River; Arthur T. Klyberg, director of the Rhode Island Historical Society; Mrs. Louise R. Hussey, librarian of the Nantucket Historical Society; Alice B. Almy, curator of the Bristol Historical Society; and Richard C. Kugler, curator of the Old Dartmouth Historical Society and Whaling Museum. For their courtesy in answering queries and providing information I am grateful to the staffs of the Wheaton College Library, the National Archives, the Peabody Museum and Essex Institute of Salem, the Nicholson Collection of the Providence Public Library, the Newport Historical Society, the Mariners Museum of Newport News, the Baker Library of the Harvard Business School, the University of Florida Library, the New York Public Library, the East Greenwich Free Library, the Varnum House and Museum, the Historical Society of Pennsylvania, the City Historical Society of Philadelphia, the Lawrenceville School, the Newberry Library, the Museum of the American Indian, the Dukes County Historical Society, the Suffolk County Whaling Museum, the Smithsonian Institution, and the Library of Congress. Mrs. Carola Paine Wormser of Alfred W. Paine, Booksellers, kindly provided information on copies of *"There She*

Blows" sold; Stuart C. Sherman, author of *The Voice of the Whaleman* and in charge of the Morse Collection at Brown University, consulted in my behalf his private catalogue of whaling logs; Hennig Cohen of the University of Pennsylvania looked up for me the book's publisher.

Many other friends have also contributed help and encouragement, among them especially Katherine Burton of Wheaton College, Norman Holmes Pearson of Yale University, John Seelye of the University of Connecticut, Howard Webb, Jr., of Southern Illinois University (who arranged a pleasant lecture in Carbondale), and Mr. and Mrs. Herbert DuBois and Mr. and Mrs. Sebastian Small of Norton. Julia Curtis Dahl cut pages. Wheaton College generously furnished me a grant so that I could travel to Hannibal, where Mr. and Mrs. Ely cordially showed me the prairies of West Ely celebrated in their grandfather's writings and answered many questions. Marion Brewington of the Kendall Whaling Museum has been a major and constant source of help. Editing this book by my great-grandfather has been fun, and the pleasantest part has been my association with people interested in whaling. They are good people all.

C. D.

Norton, Massachusetts
September, 1970

Introduction

"LOOKING back over a period of seventy-nine years," Ben Ezra Stiles Ely wrote in his manuscript autobiography, "it seems as though my life had been a kaleidoscope, changing, changing, and producing at every turn the strange and the unexpected." Sailor, pioneer, lawyer, legislator, miner, art dealer, minister — the man who wrote at the age of nineteen or twenty the history of his adventures on the whaling bark *Emigrant* had indeed a rich and varied life. But though he lived to an advanced age, no part of it was so full of incident and color and the excitement of the far-off as the twenty-seven-months' voyage that he narrated in his little book *"There She Blows."* And though much of his autobiography is told with verve and style, none of it is so carefully and successfully fashioned for literary effect. *"There She Blows:" A Narrative of a Whaling Voyage, in the Indian and South Atlantic Oceans* can well take its place among those fascinating accounts of sea voyaging and whaling that make American maritime literature some of the best in the world. It is also interesting because so much can be found among contemporary documents and records to substantiate and flesh out Ely's published account. Two of those corroborative accounts — the journals of Charles F. Tucker and Seth F. Lincoln, shipmates of Ely aboard the *Emigrant* — are included in this volume as Appendixes III and IV. They provide revealing contrasts to the picture drawn in *"There She Blows."*

Ely's antecedents were not those that would lead one to believe that he would have successfully taken his place as a foremast hand in the dirty and dangerous business of whaling. Nor would his relatively short stature — 5 feet 5½ inches, with dark complexion and black hair — indicate the physique that one might expect in a sailor. But even from his earliest days he seems to have been a vigorous and daring boy, loving excitement and danger.

> I was born [his autobiography begins] on the fourth day of February, 1828 in the city of Philadelphia in the old colonial mansion known in the history of that city as the Waln House at 144 South Second Street. My father was the Rev. Ezra Stiles Ely, D.D., pastor of the old historic Pine Street Presbyterian Church. My ancestors on my father's side were for not less than seven generations believers in the faith held by that church and were descendants of English Presbyterians, French Huguenots and Dutch. On my mother's side, my ancestors were Scotch and Scotch-Irish Presbyterians. My grandfather, the Rev. Zebulon Ely, was for forty-one years minister of the church at Lebanon, Connecticut, and

> the intimate friend and pastor of Governor Jonathan Trumbull of Connecticut, the friend of Washington and his Secretary of State, from whom the name Brother Jonathan applied to Americans was derived because, as tradition has it, Washington was in the habit of addressing him as Brother Jonathan. In the old graveyard at Lebanon the remains of Governor Trumbull and his wife lie side by side with those of my grandfather and grandmother. They who were intimate friends in life are not separated in death.

His father and two of his uncles being Presbyterian ministers, many of Ely's early childhood experiences were centered in the church. He learned hymns at his mother's knee and preached his first sermon at the age of four to his cousins and his sisters, using the back stairs for his pulpit and a chair for his desk. He officiated at his first funeral when his pet canary bird died. "My nurse," he says, "made a shroud for it, using as a casket a seidlitz powder box. My playmates composed the funeral cortege, and as the officiating minister I buried it under the old pear tree in the back yard." But Ely even as a child was far from a pious goody-goody. He was thrilled by military glamor. At the age of six or seven he was the captain of his own military company.

> We wore [he writes] shakoes made of newspapers, cockades of red, white, and blue ribbon, and I marched proudly at the head of my valiant company carrying a sword that was about as large as myself that had been worn by my uncle — Captain Robinson — during the War of 1812.

> Two of my great temptations when a boy were to steal away from home in order that I might follow a military company through the streets or run to a fire. My favorite military companies were the Washington Blues, a company organized during the Revolution, and the Philadelphia Grays. As in order to get to the front gate I had to pass by my father's study window, fearing that he might intercept me, I would pull off my shoes and run in my stocking feet until I reached the street. Much to the anxiety of the family, I would sometimes be gone for hours. On one occasion when I ran away from home, my parents were so much alarmed that they called out the bellman, an official who passed through the streets in those days ringing his bell and crying out, "Lost child! Lost child!" When discovered I was found on an oyster-smack at the Delaware Wharf making a speech to the oystermen.

Ben's love of action and adventure was given even greater scope when in 1835 his father, "under the impression that it was his duty to establish a Presbyterian college and city in the central West . . . resigned his long and pleasant pastorate over the Pine Street Church in Philadelphia and removed to Marion County, Missouri," where he was instrumental in establishing Marion College and founding the settlements of Ely, West Ely, New York, and Philadelphia just outside of Hannibal. The life was ideal for an adventurous boy of seven.

> When we landed at Hannibal [his account goes] there was but one log dwelling house, owned by a Mr. Nash, who welcomed us with true Southern hospitality, and a warehouse on the riverbank. At

that time the country was but sparsely settled, and one might ride for miles through the primeval forests and over the wide, rolling prairies covered with tall prairie grass, flowers, and in the season so plentiful of dewberries and strawberries so numerous that sometimes in riding over them the hoofs of the horses would be stained red. I have seen the deer in droves, coming up from the forests in a single file from their watering places or licks, and at night sitting in the doorway I have heard the prairie wolves howling around the house. West of the town of West Ely, where our house was located, what was known as the Grand Prairie stretched away for thirty miles with no house or sign of inhabitants. To mark the roadway leading to the next settlement a single furrow was made [*text confused*] by a prairie plough to the first settlement, lest travelers who sought to cross it might be lost in the tall grass. In traveling the broad expanse it seemed as though one were at sea out of sight of land. Lands that at that time could be entered at the U.S. land office for one dollar and a quarter per acre are now [1907] worth from forty-five to one hundred dollars per acre.

Sometimes these vast prairies became a sea of fire, lighted perhaps by some careless hunter or immigrant. Swept onward by the wind that had [un]obstructed sway, the flames leaping upward toward the skies, until it seemed as though the very heavens were aflame.

When not more than twelve years old, I became an ardent huntsman, seldom returning from a still hunt, without game. When following the hounds on the trail of deer or wolves I had the reputation of being a fearless rider, leaping over ravines and sometimes riding down the side of almost precipitous bluffs, so steep indeed that my horse would slide down upon his haunches. The excitement of the chase banished

> fear. On one occasion when following the hounds, at full run, my horse's feet struck a gopher hill and she turned a complete summersault crushing my limb. Upon killing my first deer I was wild with joy. When the hounds were on its trail, the deer would run through certain places known as stands at which hunters were stationed to shoot them. On the hunt to which I refer, as there were not men enough to fill all the stands I was given a long-barrelled old-fashioned gun that I was scarcely able to raise to my shoulder, and took my place at one of the stands. Sitting there upon a log in the woods listening to hear the cry of the hounds when they struck the trail, I heard a step — tip — tip — tip — among the dry leaves and turning my head I saw a young buck with his head protruding from the underbrush. I was suddenly seized, as I jumped up and tried to level my gun at him, with the buck ague, my knees and hands trembling with excitement. Somehow or another between my jerks I got my gun pointed, pulled the trigger and the buck fell. As he did so, I shouted out at the top of my voice, "I've got him! I've got him!" When we carried the deer to the house, the older huntsmen told me that, according to a custom among the Indians, when a young brave killed his first buck he was entitled to have a squaw.

Though he himself lived in comfort on the thousands of acres his father had bought at West Ely, young Ben got to know all the assorted characters of the frontier. There was the pioneer or backwoodsman, "generally long and lank, clothed in Kentucky jeans or buckskin, wearing a coonskin cap or slouched hat. Over his shoulders were slung a powder horn and bullet pouch, around his waist a belt with sheath or bowie knife, and carrying with him as his constant companion a long-barrelled Kentucky rifle." There was, in contrast, the Virginia, Kentucky, or Tennessee gentleman, who "could not be surpassed" for "chivalry, hospitality, and charming *suaviter in modo*." There were the ministers who preached at the revivals and camp meetings and who took an active part in organizing the Underground Railway, then so bitterly controversial an activity of the region. There were slaves, and, lowest of all, there was the Negro-trader, than whom, Ely writes, "no character was more thoroughly despised [even] among Southern people." Indeed, Ben had reason to remember that character well, since during his childhood one slave-dealer tried to kidnap Dr. Ely's only slave and was prevented by Ben's invalid mother arising to ride eight miles on horseback through the night to summon a posse.

After the death of Ben's mother, Dr. Ely sent him East so that he could attend school, and Dr. Ely, who had lost several fortunes in frantic land speculation at West Ely and later in the celebrated Marion City debacle, seems to have followed soon after. In 1842 or 1843 Ben became a pupil at the Lawrenceville School in Lawrenceville, New Jersey. But he was soon dismissed in disgrace. The story is an interesting one not only because of its revelance of Ely's somewhat scapegrace personality but because of the light it throws on the rhetoric and standards of the day. In Ely's own words:

> I remained at Lawrenceville less than a year, and

the occasion of my leaving there has always been a matter of regret. It was the result of a difficulty I had with two of my schoolmates. Up the road from our school there was a female seminary of which Miss Gulden was Principal, and some of the boys, myself among the number, had our favorite girls there. The boys slept in dormitories and not in separate rooms. One night as we were lying abed one of the boys who wished to pick a quarrel with me shouted out with an oath, "Ely's so and so," mentioning the name of my sweetheart, "is a———," using a vile name. In the flash of a moment every drop of blood in my veins was on fire and, leaping from my bed, I gave him a thrashing. The next morning as I was standing on the campus near a very low fence, a cousin of the boy I had punished came up behind me and striking me on the back of my head knocked me down. Jumping to my feet I caught up a sharp pointed picket that happened to be lying on the ground and thrust it into him, making a slight wound. The result was that I was summoned to come to Dr. William Hamill's room. In answer to his very kind inquiries about the affair, I gave a truthful account of the circumstances and said, "Dr. Hamill, I am sorry that I have transgressed the rules of the School, but under the same circumstance I would do the same thing again." He replied, "Don't you know that our Savior said that if you are smitten upon one cheek you ought to turn the other?" I answered, "Yes — but I do not believe that." He asked, "You do not mean to say you do not believe the Bible, do you?" I said yes. After talking with me in a very kind and loving manner for some time, he asked me if he should pray for me. I replied I was very much obliged to him for his kindness, and that I did not have any objections, though I did not think it would do me any good. [On my] refusing to kneel, the Doctor knelt down and offered

> a very tender and loving prayer on my behalf.
>
> After some deliberation he finally told me that I had been a good pupil but that in justice to the boys who had been committed to his care and their parents, he did not think he ought to allow a boy who entertained the opinions that I had to remain in the school. In view of my good behavior and the respect and affection he had for my father he did not want to expel me — that he should suspend the boys who were involved in the difficulty, but that he would give me a letter to Delaware College, at Newark, Delaware.

Note that the boy was dismissed not for fighting but for his "opinions," for not believing literally every word in the Bible.

Dismissed from Lawrenceville School, Ely entered the preparatory department (probably Newark Academy) of Delaware College, now the University of Delaware. But again his academic career was not long. After only a year there, "for reasons," as he says, "that it is not necessary to mention," he went to live at the home of Mrs. McClellan, his new stepmother's mother, at East Greenwich, Rhode Island, where he continued his studies under a tutor. One wonders about the mysterious "reasons" for which he was rusticated. Was he, as he says in *"There She Blows,"* in ill health, or had he been guilty of some further escapade involving the fair sex? At any rate, to East Greenwich he came, lived happily some months, and there became so intrigued with sailing and the sea that he yearned to ship aboard a vessel bound to the ends of the earth. As it had on so many young men of good family, Richard Henry Dana's *Two Years before the Mast,* published only three years previously, had a strong effect on Ely. A-seafaring he would go. But his own account, which has striking parallels with Melville's semifictional version of the same situation in the first chapters of *Redburn* (1849), is better than any paraphrase:

> The McClellan mansion [now the General Varnum House Museum] was an ideal home. It stood upon the brow of a hill, under the shade of four grand old elm trees, looking out upon the town beneath and over Narragansett Bay. It was a historic home dating back beyond the Revolution. The room I occupied had once been the guest chamber of Lafayette and Count D'Estaing and was the residence of General Varnum. There was a glamor of the past that shadowed the old house. I remember that there was a secret stairway beside the chimney which led from the dining room to the chamber above, the entrance to which was concealed in the paneling of the wall. The house and its surroundings were in harmony with the "lady of the Old School" who was its mistress and the three lovely daughters, whose refinement and lovely character filled the home with sunshine. If I should yield to the promptings of my heart I might fill pages in recording the pleasant memories of my Greenwich home and its pleasant incidents. My favorite amusement and recreation was fishing and yachting on the Bay. I became quite a skillful and venturesome sailor, taking more pleasure in strong winds and rough seas than in a calm. I had read Dana's *Two Years Before the Mast* and learned to sing "A Life on the Ocean Wave," "A Wet Sheet and a Flowing Sea," "Britannia Rules the Waves," etc. These things and a restless, excitement-

The General Varnum House, East Greenwich, Rhode Island, in an old photograph. Ben Ely was living here when he signed on the *Emigrant*. (Courtesy the Rhode Island Historical Society)

loving disposition will in part account for my longing for the life of a sailor which finally led me to leave my home to dare the dangers of the deep.

My first choice was to secure a position in the Navy as Midshipman. To this my father would not consent, fearing that the very general dissipation that then existed among the officers of the Navy might exert an unfavorable influence upon me. I then tried to get a berth on a merchantman, but failing in this I determined to ship on a whaleship. Sixty years ago New Bedford, Massachusetts was the great port of the U.S. whalefishery, sending out, when this industry was at it best, 410 whaleships. Very naturally therefore I visited New Bedford to find a ship.

I visited several shipping offices, hoping to obtain a berth, without success. It had never occurred to me that my personal appearance was not very sailor-like. I was somewhat particular in the matter of personal attire and was quite favorably [fashionably?] dressed. The consequence was that looking at me quizzically the proprietors of the several offices questioned me in a jocose manner. One asked, "Young man what has been your occupation?" I replied that I had been a student. "O!" he said, "Maybe you are a sea-lawyer. We don't want your kind." Another said, "Young man, are you in love?" A third asked me if I had a stepmother. I returned to East Greenwich quite chop-fallen but determined that I would profit by my experience. So learning that there was a whaling vessel fitting out at Bristol, just across the bay from East Greenwich, I put on a sailor's red shirt, loose pantaloons and hat, and, crossing the bay in my sailboat, succeeded in shipping before the mast on the Bark *Emigrant* of Bristol, Rhode Island.

Thus when on November 10, 1844, the *Emigrant* left Bristol harbor and sailed down Narragansett Bay for the Indian Ocean, the greenest of the greenhorns on board was Ben Ezra Stiles Ely, son of the Reverend Ezra Stiles Ely, Doctor of Divinity, sometime Stated Clerk of the Presbyterian Church, member of the kitchen cabinet of Andrew Jackson, professor at Marion College, missionary to fallen women in New York (and father of a celebrated courtesan), cogent voice in theological controversies. The little book *"There She Blows,"* published in Philadelphia in 1849, is the record of Ben Ezra's adventures and impressions. Perhaps in a manner it is a record of his growth to maturity too, though it makes no pretense (as Melville's fictional *Redburn* and *Moby-Dick* do) to be an account of social or spiritual "education."

Ben Ely was at sea twenty-seven months. But his adventures did not end when on a cold, snowy February 3, 1847, he left the *Emigrant* tied up in Bristol and walked most of the way to East Greenwich. Indeed, he was even then two days short of nineteen, and most of what he long afterward modestly called "a somewhat eventful" life was before him. After the welcome at East Greenwich described in the book, he joined his father in Philadelphia and studied law in the office of Judge Todd. But, perhaps because there was friction with his father's second wife (the children of Dr. Ely's first wife seem to have looked down on his second), he went back to West Ely to live with his boyhood friends. There, though he was still less than twenty-one, he was called to the bar as attorney and counsellor at law. On September 29, 1848, he married Miss Elizabeth Eudora McElroy, daughter of an old Kentucky family which had large landholdings near the foreclosed Ely properties. The future must have looked secure.

But something happened. Whether the McElroys' slaveholding sympathies turned them against him, or whether (as there are stories) he took to the bottle, or whether — as is most likely from the evidence — he got into a scrape with a girl after his marriage, or whether his restless nature merely demanded more action, after only a few months of married life Ben Ely was off to California as a gold miner. He landed in San Francisco in the spring of 1850. Practically broke, he first sold paintings sent out on consignment in the ship in which he had rounded the Horn to gambling saloons and whorehouses. His was, he asserted, the first art gallery in California. Next, without great success he hung out his shingle as a lawyer. Later he managed a boarding house and restaurant and even acted as bartender, though he afterward was grand master of a temperance lodge. His two or three tries at gold mining brought in some money, but were not successful for long, and once he would have died from hunger and exposure if a Good Samaritan French Canadian had not succored him. In the rough and raucous days soon after California had become the thirty-first state of the Union he was a member of the Legislature and was on the floor when a lobbyist tried to shoot a representative with a pistol. Ely nearly succeeded in

Ben Ezra Stiles Ely in the later years of what he modestly called "a somewhat eventful life." This is his only surviving photograph. (Courtesy Ben Ely)

getting himself appointed a judge. He tried practically everything, was involved in everything. But his greatest success was in persuading his able and loyal wife, who soon had regretted that she let her family persuade her to break with him, to join him in California, where for much of the time she supported the growing family by teaching school.

Even for this wayward scion, however, family tradition was strong, and the voices of generations of clerical Elys sounded loud. Harking back to the plans of his childhood and his experience of conversion on the *Emigrant* (described in his own words in Appendix I of this volume), Ely gave up his legal and political career and was ordained a Presbyterian minister. His first sermons after his ordination were preached in a theater–saloon–gambling hall (and what else?) at the mining camp of Gold Hill near Virginia City, Nevada. Thereafter he held pastorates in Healdsburg and Stockton, California; Chicago; and finally Ottumwa, Iowa. He died at the age of eighty-two in Des Moines in 1910.

Just when Ben Ely wrote *"There She Blows"* cannot be ascertained exactly. He had, he says, started a journal on board ship, but that and his letters were confiscated and destroyed by the captain for fear that they would reflect opprobrium on him. Several of the descriptions of places at which the *Emigrant* touched, such as those of Mauritius, Tristan d'Acunha, and Madagascar, have such full and exact geographical and historical detail that one suspects that Ely was writing with reference books at hand. The poems were probably the earliest-written and certainly the earliest-published portions of the book. A contemporary review pasted into the copy of "*There She Blows*" in the Morse Collection at Brown states, "The work contains a few poetical productions of the author, which have been previously printed in the Observer, and which mark him as a young man of no ordinary genius." The book was entered for copyright in the office of the clerk of the District Court of Eastern Pennsylvania in 1848, the copyright copy being deposited on October 10. It was published as of 1849 (a common practice at the end of a year) by James K. Simon, a Philadelphia bookseller. When it was written, Ely was at most twenty years old.

"THERE SHE BLOWS:"

A

NARRATIVE

OF A

WHALING VOYAGE,

IN THE INDIAN AND SOUTH ATLANTIC OCEANS;

BY BEN-EZRA STILES ELY.

PHILADELPHIA:
JAMES K. SIMON,
S. E. CORNER OF FIFTH AND SPRUCE STREETS.
1849.

"THERE SHE BLOWS."

Introduction.

THIS title, to landsmen, and those unacquainted with the business of whaling, may seem strange, but when the reader learns, that to hear this cry the young mariner first tempts the dangers of the deep; and that this is the war shout, which ushers him into battle with the lords of the ocean, it may not be regarded as an inappropriate name for this publication, in which I shall describe my voyage in the Bark Emigrant, to the Indian and South Atlantic Oceans. Like many young lads I desired to "go down to the sea in ships," that I might see its wonders, and have experience of a sailor's life. My disordered state of health, which disqualified me from prosecuting my literary pursuits, induced my father to consent to my wishes on this subject;* and now having had enough of life on the ocean wave to satisfy my curiosity and ambition, I shall attempt a faithful narrative of my past experience, in hope that it may do some good to restless sons of indulgent parents, who are anxious to leave home for rest and amusement on the world of waters. If they will go to sea, let me assure them that the hardships and dangers of a mariner's life are many, and that a whale-ship is not the abode of peace, plenty and comfort.

---

*Ely's autobiography says nothing about ill health. This may be a reminiscence of Richard Henry Dana's *Two Years Before the Mast*, the book that inspired Ely to go to sea. [All notes by the editor save as otherwise indicated.]

CHAPTER I.

Our Departure.—Sea-Sickness.—Hard Fare.
Duff.—Hunger.—Pickled Oysters.

IT was on a clear, cold morning, the tenth of November, A.D. 1844, that the rising sun had just thrown her rosy mantle over the frost gemmed earth, and sparkling waters of Narragansett, when with a "Yho! heave O!" we weighed anchor, and loosed our snow-white sails to the breeze.

It was a gallant sight to view our noble ocean bird, as she spread her white wings to the gentle gale, ploughing up many a bright gem from its wavy bed, and shaking them from her prow, as she stood boldly onward, as if conscious of her freedom.

Yet amidst all the beauty of this scene, I could not repress the unbidden tear, which, ever and anon, trickled down my cheek: for I was leaving my native land, my home and my kindred, for years, perhaps forever.

The owners of the bark in which I sailed, were

Newport Bay, where the *Emigrant* ran into trouble on both outbound and return voyages. Color lithograph, ca. 1850. (Courtesy Kendall Whaling Museum)

the Messrs. Churches, of Bristol, Rhode Island; from which place we sailed down the bay as far as Newport.

Here, the wind and tide hauling ahead, we were obliged to cast anchor, and remain for the night.

The next morning, the wind having come out fair, we weighed again, and bore out past Beavertail,† into the broad Atlantic: and long before the bright sun had sunk to rest in the Western wave, home was fast fading from our view: and more than one diamond drop of memory mingled with the blue sea, as I leaned over the rail to take my last fond look of the dear land of my birth, the home of the sailor's pride.

But the sad thoughts I indulged, were destined to be speedily superseded by feelings, which, if not mentally as unpleasant, were certainly in a physical point of view more so: for I can assure the reader

> ——"He who has never tried
> And danced in triumph o'er the waters wide,"

can have very little idea of

> "The pulse's maddening play,
> Which thrills the wand'rer of the trackless way,"‡

---

†A rocky bluff on Conanicut Island in Jamestown, across the bay from Newport. A lighthouse and a fort also bore that name.

‡Byron, *The Corsair*, Canto I, lines 13–16:
Oh, who can tell, save he whose heart hath tried,
And danced in triumph o'er the waters wide,
The exulting sense, the pulse's maddening play,
That thrills the wanderer of that trackless way?

after having his stomach turned completely topsy-turvey, and straining for hours to make it give forth the food which it had rejected hours before.

I can assure the reader, on the authority of three weeks of thorough experience, that there is no better cure in the world for "the horrors," than a good fit of sea-sickness.—Had I been such a philosopher as Goldsmith's galley slave, so that I could have looked upon others with indifference to my own sickness,§ I should, I think, for a time at least, have felt relieved from my melancholy, by the ludicrous appearance of those who were suffering around me; for at every roll and pitch of the ship, some poor fellows bore testimony to their effects, "by casting up their accounts."

It contributed nothing to the comfort of our situation, that many of us, notwithstanding our nausea and want of experience in the management of a ship, were obliged to perform duty aloft; and more than one cascade, not of crystal waters, was seen to come pattering on deck, from the masts and yards, as we were taking in sail, much to the annoyance of the officers and hands on deck.

My sickness continued violent for three weeks; and for eight months I could take scarcely any sustenance but hard bread, water and raw onions. Nothing but sheer necessity could have induced me ever to have eaten any of the other provisions which were furnished for us; for no swine that gleans the

---

§A reference to Oliver Goldsmith's translation of Jean Marteilhe, *The Memoirs of a Protestant, Condemned to the Galleys of France for His Religion.*

gutters ever subsisted on viler meat and bread than did our crew.

It may seem incredible, but it is nevertheless true, that our beef and pork in general would produce a stench from the stem to the stern of the vessel, whenever a barrel was opened. It was old and partially decayed meat. Much of it was green and putrid. Not a little of our bread was so infested with vermin, that after having crumbled it into hot water and molasses, which was called coffee, I could skim the worms and weavels off with my spoon.

Once or twice a week we had something supposed to be a flour pudding, which the sailors called *duff*. It was made by mixing together flour and water, one-half fresh and one-half salt, with a little of the liquor skimmed from the pot in which our offensive meat had been boiled. This compound was then put into a sail bag, and boiled in sea-water. This was a great luxury to half-starved men; and I have made many a hearty dinner out of a slice of this stuff, when I could obtain a little raw marrow from a beef bone for butter.

The quality of our food was not the worst thing we had to complain of; for had we been able to obtain a sufficiency of it, bad as it was, we might have become, in the course of time, comparatively speaking, contented; but we could not: and I have frequently seen the hour in which I would eat the meat raw; having procured it without the knowledge of our captain.

The way in which I managed to procure it was

this. When the officers were below, at their meals, I would watch my opportunity and steal off to the *harness cask,* in which the meat was kept, and cut myself a piece with my jack knife. I would then scrape the green putrid fat off, and eat it with a piece of wormy ship biscuit.

Sometimes we desired food at night, while standing our watch on deck. At these seasons it was more difficult to obtain it, because some one of the officers was commonly in sight, and frequently more than one. The reader would be amazed could he know the means used by the crew to gratify the cravings of appetite.

The chief of the watch to which I belonged, would wink at the stealth of his gang; but the starboard watch had many difficulties to encounter.

I remember one night, that an Irishman, called Mickey,|| wanted some pork to make "pickled oysters." This dish was made by cutting raw pork up fine, and covering it with pepper sauce and black pepper.

Mickey was at a non-plus, because the second mate was on deck, and sitting beside the "harness cask." In all such cases of emergency we had one resource; which was the ingenuity of an old canvass backed "man-of-war's-man;" who was a true Yankee tar, and could always manage to "boxhaul" in a strait. We had not seen whales for a long time, and he knew, that it being a bright moon-light night, every

||Matthew Gill. For information on the officers and members of the crew, see Appendix II.

one would be anxious to see one. So he told Mickey to go on the weather-bow, and sing out, "there she blows!"

Mickey walked to the windward, and after standing there a few minutes, cried out "there she blows! there she blows!"

"Where away?" bawled the mate, running forward.

"On the larboard bows, sir. Don't you see him?"

In this manner Mickey made out to detain him, until Old Joe# had thrown a shoulder of pork down the forecastle; which of course, lasted for some time to make "pickled oysters."

#Joseph Knights, aged thirty-four.

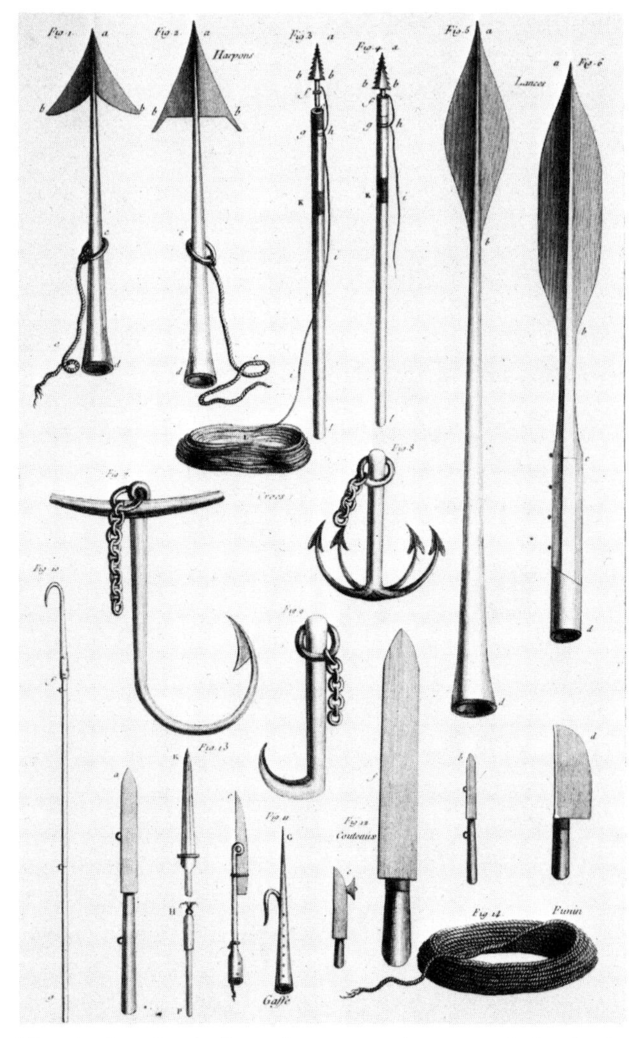

The tools of whaling—harpoons, lances, whale hook, etc.—had a grace and beauty evident in this old French engraving. (Courtesy Kendall Whaling Museum)

CHAPTER II.

Description of the Captain, Officers & Crew. A Tribute of Respect to the Second Mate.— My whole Wardrobe. The Cook and his Pig.

IT may scarcely seem credible to some of my readers, that any human being should be so lost to every feeling of humanity as to attempt to starve a set of workingmen, for a little paltry gain, as did the Captain of the Emigrant. It would be impossible for me, should I attempt it, to give my readers adequate ideas of the villany which is often practised by the owners and officers of whale ships. I can give them a sample of one in Captain James S——* of Fall River, Massachusetts, whose meanness and oppression should be published to the world, as a proper punishment of his misconduct, and to warn young men to look out well before they enlist themselves for a long voyage.

---

*Shearman. See Appendix II.

I have seen him watch his mate to ascertain how much butter he ate; and I was told, by the steward, that he actually poured sperm oil into the butter keg, so that the officers could not eat it, and that he might be able to affirm that he had butter on hand.

The first morning after we had lost sight of land, we were all called aft to receive our instructions in a speech from Captain S——. It was, as nearly as I can recollect, to this effect, and in these words: "Now boys, you've come aboard this ship of your own accord; and I'll be d——d if you haven't got to obey orders. If you don't, you'll find that I am some. I'll be d——d if you don't by G—. I won't have any of your d——d grumbling; but when I tell you to do a thing, I want you to do it. If you don't, I'll heave hell at you."

A part of his oaths I have omitted; and would have cancelled all of them, could I have given a faithful sketch of his speech without them. To say that he was always abominably profane in speech, would be uttering the truth in a very diluted state.

It is strange that such a creature should exist upon earth; but many poor Tars can testify that many such incarnate devils are in command of ships. I write none of these things because of any enmity against these persons with whom I sailed; but in justice to my fellow mariners; and with the hope it may aid in making some reform in the whaling service. I do not mean to intimate that all the faults of our Captain were natural to him, for they were acquired: nor do I represent him as being worse than many other persons would be, were they educated as he has been, to cruelty and iniquity.

To show how vile and detestable a man may become, when free from all the restraints of society and religion, I will give a few more specimens of the inhumanity and villany practised on board many whale ships. I say *many,* because there are a few exceptions to the general rule, that whale ships are, more emphatically than even ships of war, "floating hells."

It was customary on board our bark to have beans boiled once or twice a week, for all hands. On such occasions, we who were foremast men, generally received nothing but the water and a few of the refuse beans. Not thinking this sufficient, one day we sent aft, to procure some bread to crumble into our bean liquor. This threw our Captain into a violent rage, who ascended the gangway, exclaiming, "they want bread on bean day, the d——d hogs, do they?"

He was, without exception, the most foul-mouthed and degraded man I ever knew. He had respect neither for religion, nor morality, nor his own word, nor common decency. The names of mother and sister were not sufficiently sacred in his estimation, to prevent his lascivious remarks upon them. I have heard him swear one continuous string of oaths, until he was hoarse with speaking; and although he would descend to the lowest black-guardism with the sailors, yet he would treat them like dogs. In the boats, in pursuit of whales, I have heard him say "pull ahead! pull ahead! or I'll dart this lance through your d——d heart."

The young seaman has not only to submit himself to such treatment as this while on board his vessel; but before he embarks, must suffer from the shark-like propensities of outfitters and owners.

Generally, they will take a green country boy, who has neither friends nor money, and furnish him with a parcel of clothing of the most flimsy materials, at double price. The articles supplied to him in many instances, the youth cannot see until he has set sail; and therefore knows not that his outfit of apparel will soon be exhausted. To provide for this fore-ordained contingency, the owners send out several casks of clothing, of the most miserable quality, which they sell to the crew when compelled to buy or go naked, at most exorbitant prices.† By this species of robbery the poor whaleman often finds, that after having lived worse than most dogs do, for a year or two, he returns to his home shirtless, friendless, and in debt, after all the profits of his voyage have been absorbed by the sponge of the proprietors, who live at ease and sleep on their downy beds, at the expense of his toil.

Our ship's company were twenty-three in number. The officers were a Captain, a Mate, a second Mate, a Steward, three Boat-Steerers, and last, but

†See in Appendix II the accounts showing what the ship's cooper, Charles F. Tucker, bought from the ship's stores on this voyage.

not by any means least, the Cook. Of the character of the Captain my readers must already have formed their opinion.

Our mate, Samuel P. Allen, a man of naturally good temper, and had he been educated anywhere except in a whale ship, might have been a very worthy man. Had it not been for some traits of kindness and generosity in him, our company would have fared much worse than they did. He really deserves much praise for doing as well as he could under a brutal superior, and if there is any young fool, such as I was, who wishes to go on a voyage to the South Seas, I advise him to sail with Mr. Allen, if he shall be found in command of a ship.

Our second mate, Thomas Palmer was in grain one of nature's gentlemen. The feelings of a man were so deeply implanted in him, that even twelve years on board a whale ship could not wholly eradicate them. He had become in this school, however, exceedingly irascible. At one moment he would be ready to strike you without cause, and in the next, in a noble and sailor-like manner, would make confession of his fault.

Had it not been for his generosity I should have suffered extremely from want of clothing, when coming on the coast in bitter winter weather.

As a tribute of respect to him, I hereby express my most grateful thanks for his kindness, in hope this page may some day meet his eye.

Not having obtained oil on our cruise, as we expected, we were detained at sea seven months beyond the time for which we had shipped; and consequently both officers and hands, with the exception of the captain and mate, were nearly destitute of every kind of clothing. I had but one suit of garments left in the vessel. To make my last pair of pantaloons, I had cut up my last blanket, and was compelled to get one of my ship-mates to sleep with me for the purpose of keeping warm. My whole wardrobe consisted of the said pantaloons, a pair of drawers, and an old red flannel shirt which had been so much patched that you could not see any of the original garment; with an old "south wester," or hat made of tarred canvass; a pair of stockings; and a miserable pair of shoes. The patches on the red flannel were darned with about three stitches of manilla rope strands to every square inch.

It chanced that one day, having got my pantaloons wet with the waves, I came on deck in my drawers, and the second mate, without saying a word, went to his bunk and got out a shirt and a pair of pantaloons which he gave me. They were old and worn indeed, and such as hardly any beggar in Philadelphia would accept as a gift; yet still he was actually in need of them himself; and we were under such circumstances that neither love nor money could procure things necessary for our comfort. He gave me *more* when he handed me those patched garments, than had he presented me with a hundred dollars on shore.

Our boat-steerers were one Manuel, a native of

Portugal;‡ one black Joanna man, of the race of pirates;§ and one American, named Lincoln, who was a kind hearted fellow.||

In my enumeration of officers, I must not overlook the cook, who was full of fun, slush, and flour. His name was James Sables, but in seamen's language was ever called "the Doctor." He was a stout man of about four feet five inches in height, and "large at that," as he would say. He had a little pug nose, which was so aspiring that it turned up at an angle of one hundred and thirty-five degrees from the line of his upper lip. He had white hair, bow legs, grey eyes, and yellow whiskers. To finish the picture, he always wore a little red Scotch cap, cocked on one side; and when he walked up and down the deck, with a black clay pipe in his mouth, he was followed by little Jim, his pet pig, who would run, grunting and squealing after his master.

During a part of our voyage our steward was Seth Coe;# and during another part of it this office was held by the writer.

The foremast hands were thirteen in number, of whom only one was an able seaman; all the rest

---

‡Emanuel Joseph.

§That is, a native of Malay descent from Johanna, one of the Comorro Islands north of Madagascar. He may have been the man signed on at Bristol as Joseph Hammond, or he may have been a casual picked up on Madagascar.

||Seth F. Lincoln, author of the journal printed in part as Appendix IV.

#Misprint for Cole.

of us were green, inexperienced persons. Of the raw hands, one was a son of Erin: several were Portuguese, and the rest Americans.* Of one of our crew, I shall give a more extended account.

---

*For a full Crew List, see Appendix II. On the outward voyage, if one excludes the officers, three harpooners, and steward Cole, one is left with fourteen foremast hands. For the voyage in, since five hands deserted and only four new men were signed on, the figure is right.

CHAPTER III.

Old Joe,—Old Nick,—and Mickey.

ONE of the most remarkable personages on board the Emigrant was OLD JOE. He had been a sailor on board merchant vessels, a man-of-war, slave-ships, and a variety of whaling craft, and had visited almost every part of the world.

When sober he was a most interesting companion, full of anecdote and wit.

From his excellent education and polite manners, I knew that he must have been of some good family; and therefore I had a great deal of curiosity to learn his true history.

For a long time he evaded every endeavor of mine to know who he was and whence he came; but it happened one bright moonlight night, while we were sitting on the windlass, that in talking of home in a friendly manner, I mentioned the name of Mr. Emmerson,† who had delivered popular lectures on several philosophical subjects in New England.

"Emmerson? Emmerson?" said Old Joe, turning to me, "B—— Emmerson? He was a schoolmate of mine, and I was considered to be a better scholar than he was. I have helped him to get his lesson many a time, and now I am a poor d——d tar, and he is a great man! Well, if Old Joe had not been such a fool, he might have been a great and wealthy man too."

This circumstance led into a conversation, in which he incidentally mentioned the name of Miss S—— F——.

"Was she a daughter," I asked, "of Israel F——; and had she a niece named after her?"

"Yes," said old Joe, and then he told me how she was related to his father and himself. Thus he was induced to give his own personal biography.

Before I give his story, my readers may like a description of the hero of our ship.

Old Joe was a man of about thirty-five years of age. He had a dark, keen, piercing eye. His cheek was brown and withered, for many a south-wester had expended its force upon it. He appeared much

†Perhaps George Barrell Emerson, naturalist, educator, and lecturer, born in Wells, Maine, in 1797. If so, Joseph Knights falsified his age, which he gave as thirty-four, and his birthplace, which he gave as Boston, on the Crew List. But the story sounds remarkably like something out of a dime novel or temperance tract. Probably Old Joe was just spinning a yarn to greenhorn Ely.

older than he was, for deep dissipation as well as hardship had contributed to the furrows on his brow.

He commonly wore an old blue man-of-war jacket, patched all over with pieces of cloth of different sizes and colours, which he had obtained in different ports of Europe, Asia, Africa and America. It was furnished with toggels, or pieces of wood tied to strings, for buttons. He wore a "southwester," a covering for the head made of tarred canvass; and pantaloons of the same materials. He was as erect as an arrow, and as spry as a tiger. He was the most profane man among all the profane sailors, with the exception of the captain, I ever heard open mouth. Nevertheless, there were many little indications given by him, thoughtlessly, that he had been well educated; and that in his lowest degradation amiable feelings would sometimes prevail. He was not all sot; not all fiend, at all times. There were moments when latent sparks of fire would be discovered under an icy exterior: and his bosom, for the most part resembling a rock, would occasionally heave like his much loved ocean.

I will no longer detain the reader from hearing Old Joe himself.

"Well, Ben, since you must know it, I suppose I must tell you.

"I was born in the town of———, in the State of Massachusetts. My father was a rich, plain old farmer, who sent me to school until I was sixteen years old. Then he died, and I was obliged to take care of the farm with my step-mother. I lived very contentedly with her until I became acquainted with a girl about three years older than myself.

"Aye, boy, I tell you, she was a trim craft. Every thing about her was *shipshape, and brister fashion.* She'd make you open your weather eye, if you could see her. When I think of her, it almost makes my old eyes spring a leak.

"Well, I got to like her very well, but I was one of your bashful sort of fellows then; so that sometimes I used to be so 'fraid my folks would twig me when I had been to see her, that I would go four miles round, so as to come home the other way.

"Well, after awhile I wanted to marry her, and the people found it out, and used to make fun of me, and tell me I was going to get a girl who would be a mother and wife both to me. That made me feel so bad, that I went away after I had married her secretly; and here I am a poor d——d tar that no one cares for."

"Well, Joe," said I, "what is the reason that you don't go home to her again?"

"Ah! she's dead now," he responded, "and I shall never go home again. She was the only one I ever loved, and I could not live there now. I have been to see the boy she left me; but he don't know his father is living. All my friends, except one old uncle, in whose hands I have placed my property, for the use of the boy, think that I am dead.

"They don't know that the boy is my son, and it will make a mighty shivering amongst them, when my boy comes of age, and they find that the property they live on belongs to him.

"I've been many a long knot since then, and might have been a retired captain long before this, if I had not been such a d——d drunken fool: but when I get on shore, I know no one cares for Old Joe, and that even my boy would not own me, and so I think I might as well be jolly as sober."

Before going to sea, I had read some of the gloomy misanthropic poets, who from a false view of life, combined with disappointment, look upon mankind as wholly a deceitful and selfish race. Unfortunately, I too, had become despondent, and my romantic authors led me to look for and expect perfect friendship or total depravity in all with whom I should have intercourse. Disappointed in finding disinterested and pure benevolence, I adopted the Byronian views, and was disposed to attribute every action of every man to self-love.

One would imagine, that in being thrown, as I was, on going to sea, into the company of low and despicable miscreants, my gloomy misanthropic sentiments and feelings would have gained strength. But directly the contrary was the case with me; and I learned, that however base and degraded human beings may become, there are still some principles slumbering in their bosom, which angels might not blush to own.

Old Joe was an example of this character. He had been a slaver, a merchantman, a man-of-war's-man, and a drunkard, and had received additional debasement from each; and yet some noble and generous impulses remained in his heart, which, like the sun, obscured by clouds, would ever and anon beam forth, to remind you that its light was not all gone.

You might well suppose that if any man on earth is destitute of all the kindly emotions of humanity, that man must be Old Joe.

We had some difficulty with the second cook, who was a disagreeable black fellow,‡ and Joe often said, "you ought to knock the —— nigger on the head, and heave him overboard, as we used to do in the slaver." Such a sentiment as this could come from no one who was not as depraved as human nature could be; nevertheless there was much in him to be loved and admired, and had the fear and love of God reigned in his heart, he would have been an ornament to society. Let me illustrate my meaning by two anecdotes concerning him. Having spent twenty-seven months of indescribable hardship at sea, he landed on the wharf at Bristol, R. I. with only thirteen dollars due him for all his hard labour. This was all he had to keep him while on shore, in food, drink and clothing. On leaving the vessel he met a little girl with scarcely any covering to her feet. He stepped up to her, and taking her kindly by the hand, asked "have'nt you got any better shoes and stockings, my little girl?" "No sir," said she; whereupon he took her into a store, gave her some cakes, and

---

‡Probably the man signed on at Port Louis as Henry Henry, the only member of the crew with black complexion and woolly hair.

left money enough to buy her a pair of shoes and of stockings.

He then went into a grog-shop, and while sitting there, a poor Scotch sailor entered, whom the land-sharks had cheated out of all his earnings, and asked the tavern-keeper to give him a shilling, for he was a stranger without a friend, and without a home. The land-shark of a publican refused; but Old Joe, putting his hand into his pocket gave him about half of the money he had left, saying, "take that, shipmate, it will do you more good than it will me, for I shall spend it for rum."

Early in the summer after my return home, this old tar came in a coal vessel, on a coasting voyage, to Richmond on the Delaware;§ and by inquiry found his way to my residence. He asked for my father and myself, who were not at home; and then sent a message up stairs to the family, that "Old Joe, one of Ben's shipmates, had come." In he bolted, in a state of intoxication, and said he could make himself perfectly at home, until I should return.

Of this he gave the *Eastern* evidence of tilting back his chair, and the *Western* evidence of spitting tobacco juice on the carpet.

Unasked he said he would stay for dinner. After some difficulty the family got rid of him; but at night he returned again, and not finding me, and the weather being warm, slept on a pile of boards, in a

§Port Richmond, a section of Philadelphia along the Delaware River.

vacant lot opposite our dwelling. Frequently in the night, in his drunken sleep, he would mutter, curse, and groan; and in the morning he was at the door again before I could dress myself.

Poor fellow, I was indeed *glad* and *sorry* to see him, for the hospitalities of no decent house could be tendered to such a sot.

He made many apologies for having intruded the day and night before, in a state of intoxication; and having furnished him with some necessary clothing and a little money, I attempted to get him on board his ship again. To accomplish this, I walked with him to Richmond, and tried to persuade him to sign the temperance pledge: but all in vain, "for no one cared for Old Joe," he said.

At every grog-shop which he passed on our way he would stop and drink; until he became so dreadfully drunken that I was obliged to leave him at an inn near his vessel. He called again at our house; but it was necessary I should refuse to see him, because he only wanted strong drink; and some weeks after he had gone, I saw the articles of clothing which I had given him when almost naked, at a pawn-broker's shop. He had parted with them to get a little grog money.

Alas! poor Joe! he had good natural talents, a good education, and agreeable manners, when sober. I could not help shedding tears when I left him, and when I refused any more to see him. He might be a blessing to his friends and the community; but he is carried out into an ocean of wo by the hell flooded tide of *intemperance.*

Old Joe was once in St. Petersburg, and passing along the streets in old shoes without any stockings, in his common state of destitution, he repeatedly cursed Old Nick, and said to a fellow tar, "if he were a gentleman he would give me a pair of new shoes."

"It so chanced," Joe says, "that the Emperor Nicholas was passing on foot, and heard him, and knowing him to be an English or an American sailor, from his lingo, gave him money enough to buy a jacket as well as stockings and shoes, saying, "learn hereafter to speak more respectfully of the Emperor of Russia."

A part of the jacket Joe still exhibits, and says that it was a fine looking, portly gentleman, whose outer garments were trimmed with ermine dyed black, who gave him the purchase money. Ever since Joe speaks well of Old Nick, and in all probability the great Emperor of all the Russias has some of the good qualities of a naturally kind heart, which we have discovered in our prince of wicked mariners.

On board our ship, Mickey, an Irishman, of whom I have already made some mention, was a humble imitation of Joe, in both his good and bad qualities. For one of his age, he was even more degraded than his model.

During nearly the whole voyage, he sought a quarrel with me because I was the son of a gentleman, and he concluded that I deemed myself on that account superior to him.

One night during our watch below, after saying every thing he could to insult me, he applied the most opprobrious epithets to some of my nearest female friends, whom he had never seen, but of whom he had heard me speak.

Exasperated in the extreme, I sprang from my bunk and struck him in the face. We fought for nearly half an hour like tigers, until covered with blood and breathless we were obliged to desist. I had rather the advantage, for which for a long time he never forgave me; and of course we were not on speaking terms, until the latter part of the voyage.

I was then out of that greatest of all comforts to a sailor, tobacco; and he heard me say I had none. From our hostile attitude, he would give me none, and yet, while I was on deck, he would go below, and put a plug or two into my chest.

Our voyage lasting six months longer than it was expected it would, I became destitute of shoes, and my foot was so small that I could not wear any pair in the slop chest.

One of the crew, whose soul was no larger than the eye of a mosquitoe, had a pair which would fit me, but I had nothing to barter for them.

Mickey, however, one day was trading with him for a hat, and would not strike the bargain, unless he would throw in the shoes, "to boot." The owner of them declined unless Mickey would give him more. I turned to them and said, "why Mic, what do you want of the shoes? you cannot wear them, they will not fit you."

"I know that," answered Mickey, "but they will fit you, Ben."

After this I think no one should say, that there is not some good thing left in the worst of men.

"Reefing topsails." Color aquatint, 1832, after a painting by W. J. Huggins. (Courtesy Kendall Whaling Museum)

CHAPTER IV.

History of the Voyage continued.—A Gale.—A Dream.—Homesickness.—Remembrance of Greenwich.—Drawing Water.—Seizing up.—The Captain and the Tom Cat.

NOTHING of any considerable interest occurred, until I had been three days out of port, when I experienced my first gale.|| The earliest intimation I had of it, was during my watch below, from which I was roused by a rough voice, crying out, "all hands ahoy! rouse out there, and show yourselves."

Immediately I sprang on deck, without hat or pantaloons, and there I saw a sight well calculated to excite terror, at least in the mind of an inexperienced sailor.

The moon had just broken through the dark and massive clouds which were hanging over our heads,

||See the entries in the Lincoln and Tucker journals for November 17–19.

and revealed to our view the decapitated appearance of our ship. — There she lay, her lee rail for most of the time under water; one boat off the cranes; her sails flapping furiously; her masts quivering, and her rigging hanging in tangled coils about the deck. Now and then, as the moon was darkened by a floating cloud, a gleam of forked lightning, followed by a long and loud peal of deafening thunder, would fill up the interval, throwing a ghastly pallor over each countenance, and showing the outline of our slender spars in relief against the dark bank of clouds.

To add to the difficulty of my situation, I was still deathly sick, but was obliged to go aloft to take in sails. Many would suppose that it would fill a young sailor with great terror to be required to run aloft thus, in the midst of a tempest, when perhaps he had never before in his life, been twenty feet above the level of the deck: but I was so sea-sick that I had the most total disregard to danger and self-preservation.

That I did not fall into the ocean, was more the effect of instinct than of any thing else; for I scarcely cared whether I held on or not. Those who stood under me were sure of some evil, for any new exertion which disarranged my stomach, would produce a most unpoetical cascade.

One of the principal amusements of my friend Old Joe, in the early part of the voyage, was to aggravate our misery, by speaking of every thing which was calculated to call to our imagination to the aid of nausea.

A Gale.

In this gale we split our foresail, carried away one of our boats, and experienced some other small damages.

The wind howled, and the waves roared, dashing over us, and heaving us upon their foamy bosom, as if they scorned to bear our burden.

The lightnings flashed, the thunders pealed, the bending masts creaked and groaned, and nature sang her fiercest war song; yet still there was a grandeur, a sublimity about the scene which filled me with admiration, and for a while bid me love the sea, and banish pain and sorrow from my heart.

> "O! who can tell, save he who's tried,
> And danced in triumph o'er waters wide,
> The exulting sense, the pulse's maddening play,
> Which thrills the wand'rer of the trackless way?
> Oh who can tell? Not thou, luxurious slave,
> Whose soul would sicken o'er the heaving wave:
> Not thou, vain lord of wantonness and ease,
> Whom slumber soothes not, pleasure cannot please."#

The next morning the gale subsided. The bright sun came forth, and the dark cloud gave place to the pure ether; when, for the first time, I saw the sailor's boast, of "the blue above, and the blue below."*

#Byron's *The Corsair* again, lines 9–16 of Canto I quoted, probably from memory, in changed order.

*Barry Cornwall, "The Sea," stanza 2:

> I'm on the sea! I'm on the sea!
> I am where I would ever be,
> With the blue above and the blue below,
> And silence wheresoe'er I go.

To one who has never seen old Ocean in his loveliest mein, the author's pen can give but little idea of its beauty. Far, far as the eye can reach, lies one boundless expanse, of the deepest, richest blue, save where here and there some spot of feathery foam, like a fleecy cloud upon the summer sky, or a cluster of sparkling diamonds, thrown into the sunlight by some unseen mermaid's hand, relieves the monotony of the scene; and seems to present in miniature the starry abode of the countless worlds on high.

After the gale we experienced nothing of novelty until we reached the Cape De Verd Islands. Our life was one dull round. In our watch on deck we knotted rope yarns, and platted sinit,† or made, and took in sail. In our watch below, we retired to our bunks, to think of home, or sleep.

Most of the green hands, for the first month or two, were subject to the nightmare; and one of the chief amusements in the dog-watch and watch on deck at night, was to listen to the relation of the dreams of persons who had laboured under that unearthly weight upon the breast.

Homesickness, to one who has never experienced it, may be regarded as a trifling thing; but there are few diseases which more entirely unman a person. In many cases it has so preyed on sailors, that it has been necessary to send them home; and some on the

†Plaited sennit or sinnet—a "flat braided cordage formed by plaiting ropeyarns or spun yarn together" (*Century Dictionary*).

way have died before they could reach their native land.

During the whole voyage I suffered more or less from this complaint, and for the first few weeks was actually confined to my bed from despondency. Had not the love of novelty prevailed and counteracted this malady, I should undoubtedly have found a grave beneath the blue seas, with none to weep over me, save the foaming billow; none to sing my funeral dirge, save the howling tempest.

I used to feel more deeply on the Sabbath than at any other time; for on board ship, there was little or no regard paid to that holy day, and I longed for the sweet solitude of my own dear home. I would sometimes steal away from my shipmates, and strive to picture to myself scenes dear to my memory. I loved to think how the church bell sounds over the village hills; how the birds sing, and the green fields smile.

The reader will, I trust, pardon me for introducing an extract from my memorandum book, which will give some correct idea of my state of mind.

REMEMBRANCE OF MY GREENWICH HOME.‡

I'm sad upon this holy day,
When far from home and friends away,
No Sabbath bell's sweet, cheerful voice,
Calls forth God's people to rejoice:
No grateful hymn ascends on high,
No fervant prayer, to pierce the sky:
No emerald field, nor shady bower,

‡East Greenwich, Rhode Island

"There She Blows"

No song of birds, no breath of flower,
Can glad the lonely Neptune's child,
While wandering o'er this trackless wild,
Where naught is seen save foaming billow,
Jove's starry couch and fleecy pillow,
Where nothing cheers my lonely heart,
Save thoughts which memory can impart.

And when alone at even tide,
I view the waters, far and wide,
With mental eyes I oft have scanned
The beauties of my native land,
Where white upon the silvery bay,
The bounding barque flies merrily.
Upon the bank, in vestal green,
The smiling fields of Kent are seen:
Far in the East, Mount Hope's green isle,
And in the West, old Greenwich smile.
There peeping shady boughs between
My mother's house to see I seem;
Its brave old elms, those hale green trees,
And garden with its fragrant breeze:
And there, in sunset's rosy hue,
Three sister forms I fondly view,
Sitting within the entry way,
And knitting till the close of day.
With gentle smile, but eye severe,
Watching o'er those to her most dear,
Their mother sits—

The winds are roaring fiercely now,
The waves are dashing o'er our prow;
But in the calm, and in the storm
The thought of home this breast must warm;
And Greenwich has a voice for me,
My heart can hear on every sea.

Drawing Water.

Such a life as a man is obliged to live on board of a whale-ship, must be unpleasant to any one, whatever may have been his condition on shore; but, to a person of any good education, and to one who has been brought up in the lap of luxury, it is painful beyond description.

Previous to my going to sea, I had never done any manual labour; and indeed I was in a very feeble state of health. Yet, notwithstanding, I was compelled to perform the duty of an able bodied man. My hands were very tender, and for weeks after I sailed they were blistered and raw, so that when I hauled on a rope I would leave the marks of my hands in blood.

The most arduous task we had to perform, was that of drawing water to wash the decks. This was done every morning. The way in which it was performed was this. One man went over the side of the vessel, with a breast mat, or rope made fast around his waist; and having caught the water in a bucket, swung it up by a rope to a hand who sat on the rail, to grasp the watery pendulum, when it had described a quadrant. I have stood in this manner and drawn water, until I have fainted in the chains, and was left to hang there until I came to my senses again. Sometimes after this violent labour for two or three days in succession, I expectorated blood; but it was in this case as in many others, "kill or cure;" and a kind Providence ordered the latter. I became so hardened at length, that I could endure as much fatigue and violent exercise as any of our crew.

There were trials of a different nature, however, much harder to be borne than these. For when I was ordered about deck like a dog, at the beck and call of men whom I looked upon as in every respect my inferiors, I found myself more than once grinding my teeth and clutching my knife. I had been accustomed to be asked, in a gentle manner, to perform any service, but when I came to hear an insignificant upstart say to me, "bear a hand you d——d lubber, or I'll give you the rope's end," it was more than human nature was willing to bear, and I fully, perhaps wickedly, determined that if I was ever seized up in the rigging, (for this is the manner of flogging at sea) either I or my tormentor would come down dead.

No slave on shore ever suffered half so much injustice as in nine cases out of ten is practised on board our whale-ships, and in many instances even on board our merchant vessels.

The inhuman manner of inflicting punishment at sea should be generally known.

The accused person, whether a real offender or not, is taken to the main rigging, where a block is made fast sufficiently high to swing him clear of the deck; a cord is made fast by a slip knot round his wrists; his feet are tied together; and he is suspended by the extremities of his arms. Then the captain, or his officer, with a whip of nine cords, each of the size of your little finger, and filled with knots, commences his *noble* operation. At every stroke he brings the blood from the naked back of his victim: and the poor fellow ever after experiencing this cruelty is either a fiend, or a broken spirited sluggard.

Methinks I hear one and another gentle reader say, "horrible, horrible!" when brine is applied to the excoriated back; but reflect, such transactions are sanctioned by our maritime laws, and these laws have been enacted and are continued in force by our legislators, and these legislators are elected by the humane people of the United States.

Both British and American seamen are subject to as horrible slavery while at sea, as any of the children of Africa in any part of the world.

The only being who was actually *seized up* during our voyage was the Tom Cat. Our manly captain seemed to derive most of his amusement from tormenting this poor animal, and the cook's pet pig.

It happened one day that my four legged friend Tom made up his mind to resist a perpetual teazing; and therefore he implanted his claws in the hand of his tormentor. This exasperated the Captain greatly, and he forthwith ordered the steward to "seize the cat up." On penalty of sharing the same fate, the steward had to obey: and the captain, with his cat-of-nine-tails whipped the poor brute almost to death. This was an honourable employment, truly, for one who considered himself a man, and who was authorized by *law* to command his ship, and torture his crew.

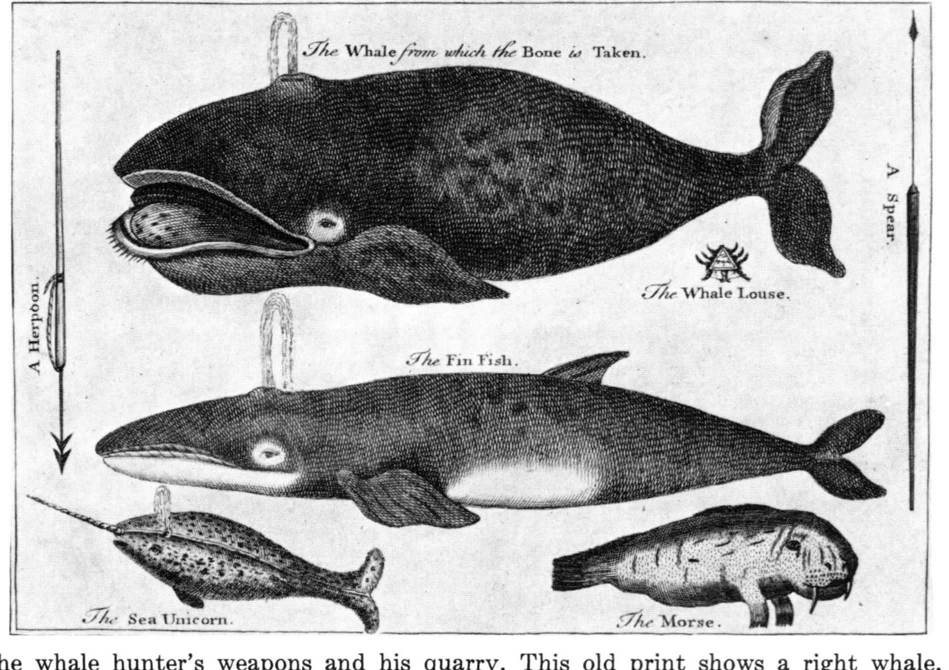

The whale hunter's weapons and his quarry. This old print shows a right whale, a finback, a whale louse, a narwhal, and a walrus, flanked by a harpoon and a lance. (Courtesy Kendall Whaling Museum)

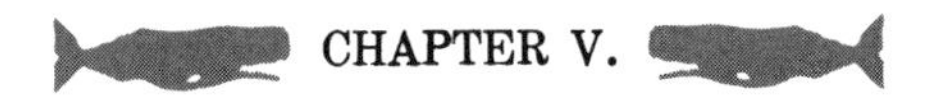

CHAPTER V.

History of the Voyage continued.—Island of Bravo.—A Witch.—Crossing the Line.—Visit from Old Neptune.—Capture of our first Whale.—A Christmas excursion at Sea.

EARLY on the morning of the 18th of December, 1844, after having spent some weeks without seeïng any thing, save our ship, our crew, the sea, the sky and its "drops of day," we were all cheered by hearing the man at masthead give the cry, ever dear to a sailor's heart, of "Land! O!" I went on deck and saw what appeared to me to be a dark blue cloud rising from the midst of the Ocean. As we drew near it began to assume a more decided form, and I soon beheld the Isle of Bravo towering like a mighty rock, high above the world of water. It bore a wild and awful appearance: rock rose upon rock for hundreds of feet, until their heights were lost amid the dark and frowning clouds, which hung in gloomy grandeur over the barren and precipitous sides of the mountain.

We could, when miles distant, hear the roar of the breakers as they dashed in mad fury against these weatherbeaten monuments of ages past, throwing their white foam for a hundred feet up their dark and rugged sides.

We stood "off and on" this Island for two or three days, for the purpose of obtaining two more hands, and a bolt for our "fore swifter,"§ so that we all had an opportunity to go ashore; and never did Trojan "lie down on the wished for sand"|| more joyfully than we hailed the privilege of placing our feet once more on "terra firma."

The natives of Bravo are a very hospitable and pleasant race of Portuguese descendants, principally of copper complexion, with fine forms, and agreeable manners. They are mostly mulatto creoles, and Roman Catholics, who believe in fairies and witches; and in their superstition are much encouraged by their priests.

Some of their witches are truly quite bewitching, with one of whom I formed a very interesting acquaintance. Instead of being an ugly old hag, she was a very beautiful young girl; and with her broken

---

§Evidently a bolt fastening to the chains the dead-eye to which was attached the swifter, a forward shroud of the lower rigging.

||Vergil, *Aeneid*, I, 171–173:

> . . . ac magno telluris amore
> egressi, optata potiuntur Troes harena
> et sale tabentis artus in litore ponunt.

English and bright glances made quite an impression on a sailor's heart. She did not seem to be bent on conquest however; for she took it for granted that I had a sweetheart at home.—She presented me with a string of green beads, and told me if I wore them, I should be safely returned to my lady in America.

The females in this Island smoke a great deal; and having motioned to me to sit beside her, she took my hand in hers, and my pipe from my mouth and made free use of it.

I told her as well as I could, how badly we were treated on board ship, and how I longed for land; at which she seemed to be much affected, and gently pressing my hand endeavoured to persuade me to leave my vessel and live with her, until I could go home. Indeed, I should have taken her advice, had a good opportunity offered, but it was ordered otherwise.

The people of this Island seemed to me to be the most happy, and in some respects the kindest I ever had seen. It is customary with them, if a stranger is left there, to take care of him for a year, and if after that time he will not provide for himself, they send him to one of the neighbouring islands.

Their favorite amusements are music, dancing, card-playing, and drinking intoxicating beverages. Very few of them appeared to be habitual drunkards.

After procuring a supply of oranges, figs, and other food, we set sail for the Line; and nothing of interest occurred until we reached that far-famed abode of Old Neptune. Here we found the weather

Crossing the Line. Neptune and his helpers shave the greenhorns in the seamen's traditional rite of initiation. Lithograph, ca. 1858. (Courtesy Kendall Whaling Museum)

excessively hot, and had it not been for the frequent showers of rain which fell, it would have been almost impossible to have remained on deck. The rail and plank of the ship were frequently so hot that we could not bear to touch them with naked hands or feet.

The burning was great even through the soles of our shoes.

Most of my readers are probably unacquainted with my friend, *Old Neptune*, and therefore I beg leave to introduce him, by describing his visit to our vessel; a thing of frequent occurrence to ships formerly, but which of late has become more rare.

On the 29th of December, at about eight o'clock in the evening of a bright moonlight night, we were all suddenly startled by hearing a deep, full, unearthly voice, crying out, immediately under our bows, "ship ahoy!" upon which our captain gave orders to haul aback the main yard, and ease down the helm. He then answered through his speaking trumpet, "Ahoy!"

Then sounded the same deep-toned voice from the stern, "What barque is that?"

"The Emigrant, of Bristol, Rhode Island," answered the captain.

"Who commands the Emigrant?" asked Old Neptune.

"Sherman!"

After this intercourse, we saw a very tall man, dressed in a long, white, flowing robe, with long black hair and beard, coming over the bows, with a tar bucket hung around his neck. He had in one hand a speaking trumpet, and in the other an immense razor of about two feet in length. After depositing his bucket and razor on the main hatch, he seized the nearest green hand, and with the help of the officers, and his half-brother Old Joe, lathered his face with tar and slush, by means of a paint brush, and then performed the operation of shaving him, with his large razor, after which he declared him to be one of Neptune's sons, of which, the poor fellow bore ample testimony the next morning, by appearing with his face and hair daubed over with tar. But as all young hands are ambitious of being called sailors, they carried their diploma of seamanship with great good humour.

On the 1st of February 1845 we took our first whale. This of course had been a long anticipated event.

A whaleman's life is one either of dull monotony, or of thrilling excitement, and of hard labour.

The danger and anxiety attending the capture of a whale, are very similar to those risks and solicitudes which huntsmen will encounter in pursuit of game. I have often felt so desirous of obtaining a whale, that I have pulled at the oar until I could not see: and yet the moment after the whale was dead, I would have rejoiced to see him sink, that I might not be obliged to perform the labour of taking care of him.

So wearisome are most of our hours at sea, however, that dangers and hardships occasionally become welcome visitors, because they change the scene.

Picture to yourself, reader, a calm and mirror-like sea, and our ship rolling to and fro on it, without the slightest breath of wind to steady her, while the sails are flopping lazily against the masts, and the hot sun is beating upon our vessel. Here and there about the decks, you see the sailors trying to jog the tardy wings of time, by courting sleep, or humming some idle hymn to Bacchus. Even the cat and pig seem to have hysterical fits from the baneful influence of inertia and ennui: when of a sudden the man at the masthead cries out in a faint and tremulous voice, as if he half doubted his own vision, "There ——— she blows!"

At his first sound, you will see some one of the idlers on deck raise his head from his arm and look aloft.

Then the man at masthead clearly sees the spouting; and with a stronger and quicker voice he says distinctly, "There she blows!" Then you will see the Captain jump up to the head of the cabin gangway, and look aloft. When the look-out sees the spout for the third time, he is confident that it is a whale, and he shouts at the top of his voice, in a clear and joyous manner, "There she blows! there she blows! A right whale, sir!"

"Where away?" asks the captain.

"Three points on the lee bow, sir!" *Citius dictu*, every man is aloft, and as the whale blows, or ripples the water, they will all sing out in joyful chorus,

"There she blows!" Whales spouting near a whaleship on the grounds. One of a series of nineteenth-century postcards. (Courtesy Kendall Whaling Museum)

"there she blows! there she blows! there she blows!" Then the captain calls to the helmsman, "keep her off three points! square in the main yard. Get your lines on the boats, and see your davy# falls clear."

After approaching as near as was practicable with the ship, the orders are given to "lay the main yard aback, and lower away the boats." As soon as the boats touch the water, every man is at his oar; and wo to him who is not; for, from the moment in which we leave the ship, until the whale is either actually dead, or spouting blood, the officers seem to be perfect maniacs. At first, they will beg you to pull, and then curse you. "Oh! do pull, my dear fellows. Do pull. Pull you d——d lubbers; pull, or I will heave this lance through your heart. Pull, my boys.—I've just got five sisters at home; and if you will get that whale, I'll give each of you a wife. Pull, I will give you my voyage for that whale." In this way they will rave during the whole time of the chase.

A whaleboat is about twenty-five feet long, and furnished with a keg of water, a lantern-keg, six oars, six harpoons, three lances, a whaleman's spade, a small waif or flag, and a line tub, containing from three to six hundred yards of line. This line is made fast to the harpoons, and while it is running out, it is necessary to keep continually wetting it, to prevent it from taking fire, by friction.

One man stands in the stern of the boat, and

#Davit.

Lancing a whale. Two boats move in for the kill in this postcard scene. (Courtesy Kendall Whaling Museum)

steers it with a long oar. The other five pull until they are within a few feet of the whale; when the officer who is steering calls to the harpooner to stand up. Then the other four lie upon their oars.

This is the point of extreme excitement. Having come within a few feet of the side of the monster the harpooner darts his barbed iron with all his might, and quick as lightning the order is yelled to "stern the boat off!" If this is not done instantly, so soon as the whale feels the dart, he is apt to crush the boat; for notwithstanding his immense size, he moves his body with the speed of thought; and sometimes will appear to stand on the very tip of his head, and to throw his flooks from eye to eye in a moment of time.

If not out of the reach of this tremendous paddle, the boat and crew are lost.

The roar of the right whale, when wounded, is terrific. I have heard it at the distance of four miles.

The capturing of a whale is certainly one of the most fearful conflicts for the six men concerned in it, which can be imagined. The whale roars and snaps with his immense jaws, and in his fury lashes the blue sea around him into bloody foam.

The men shout and the line whizzes through the chocks of the boat. Sometimes the whale disappears for a few moments, and this is the most trying time of all, for you have reason to fear that the next moment he may come up under your boat and crush it in his jaws. Such an event has often happened, and is the terror of whalers.

Frequently I have been so near a wounded whale that he has stricken my oar with his flooks.

One Christmas day* we saw whales about daylight in the morning, and finding they were going very fast, we lowered the boats at once. After pulling about ten miles, our boat got on to one; and my oar struck its flooks before the irons entered him; whereupon at the first dash, he struck three planks out of our larboard side, threw two of our men half over the lee side, upset the tub, and took the line out foul.

However, we kept the boat afloat, and were fast to him all that day, but he was so wild that we could not get an opportunity to lance the monster.

I was bowman on that occasion,† and so far as

---

*December 25, 1845. See Tucker and Lincoln. Lincoln's comment on the day was: "A fine days work for Christmas"!

†A compliment to Ely's abilities. In *Arctic Harpooner: A Voyage in the Schooner Abbie Bradford 1878–1879* (ed. Leslie Dalrymple Stair, Philadelphia: University of Pennsylvania Press, 1938, page 5) Robert Ferguson writes: "The smartest man was always made bow oarsman. He had to be fast and clever to take down the mast and sail quickly after a whale was struck, for both of those things had to be kept clear of the whaleline as it ran out a-whizzing. It was also his duty to attend the officer when he was busy lancing a whale, always keeping a lance ready for him. Furthermore, he had to haul on the line to keep the boat close up to the whale."

danger was concerned had the most honourable position in the boat.

It was my duty, when we got near the whale, to reach over the head of the boat and take the line outside of the chocks, or hole through which it was to run, and haul the boat on to the whale, while the boat-steerer laid against me with his oar, with intention to make the boat run up abreast of the whale. This would give the captain opportunity to use his lance. In doing this I was jerked over the bow of the boat, all except my knees, fifteen times during the day. Once, in the confusion of the moment, I mistook the order, and stretched myself over at the wrong time, so that the whale brandished his tail over me, and my head scarcely escaped his deathful boxing of the ears. Having been fast to this whale all day, we were reluctantly compelled to cut loose of him at night; for he had towed us out of sight of our ship, the sky gave evident tokens of a coming storm, and we expected to be lost.

My companions and myself were so much exhausted and overcome with this hard day's labour, beneath a burning sun, without any food or drink, (for the after oarsman had forgotten to fill the keg,) that we seemed almost to disregard our situation. One or two of the men fell down, almost inanimate, into the bottom of the boat. Our tongues were black, and swollen for want of water, and some of our men actually had recourse to a most revolting expedient. We could not tell the direction in which we had last seen the ship, and our captain told us we must pull for her, or be lost. I was so worn out, that I replied to him, that "I did not care, and would not, and could not, pull any harder."

Nevertheless, after toiling for about three hours longer we got on board. The relief of a little water who can describe? The other two boats were nowhere to be seen.

But having some idea of the direction in which they might be found, we stood for that point of the compass, when by mere chance, as sailors say, we discovered them on the lee bow. There these poor fellows were, all lying on the bottom of the boat, and expecting to die.

Just picture to yourself, reader, if you can, a little crew in an open boat, in the midst of the ocean, without provisions, without compass, at night, while dark clouds roll up to the windward, and portend inevitable death. This is one of the excitements and vicissitudes of a whaling voyage.

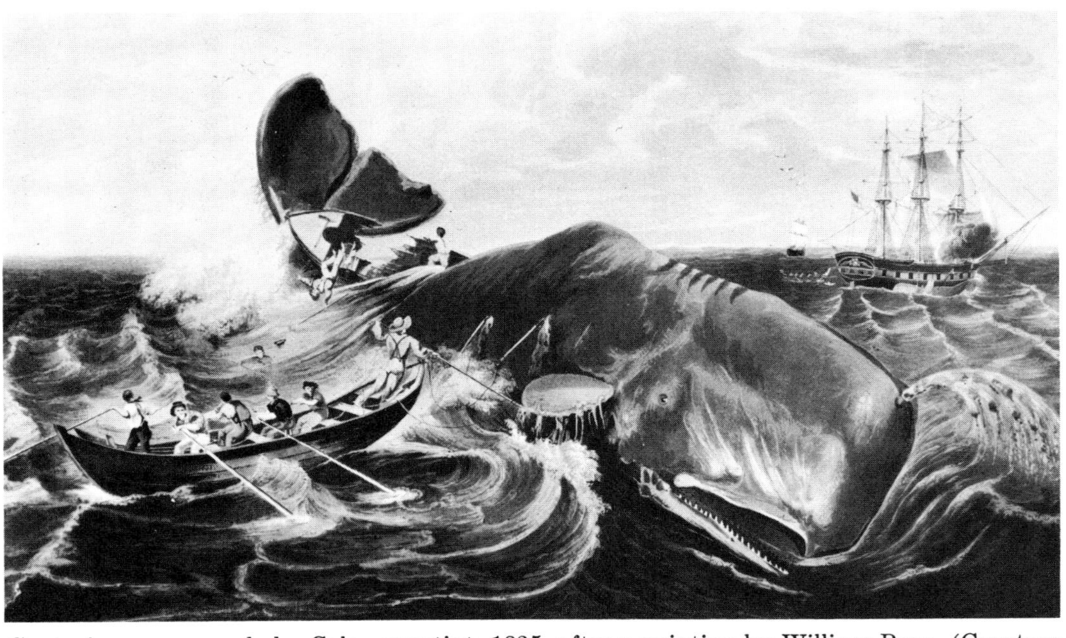

Capturing a sperm whale. Color aquatint, 1835, after a painting by William Page. (Courtesy Kendall Whaling Museum)

The death flurry, a moment of extreme danger for the attacking boat, as depicted on a postcard. (Courtesy Kendall Whaling Museum)

CHAPTER VI.

Hard Labour.—Drowsiness.—
Reeling off Blubber.—
Work in the hold.—Trying the Whale.—
New Steward.—Private letters.—
The Captain's Temperance and Tyranny.

ONE who has never experienced the hardships and dangers of a whaleman's life, can form but a faint conception of them. Indeed, were I to describe some things connected with it, and their occupation, my readers would say, "are they possible?"

I have left the ship at ten o'clock in the morning, and rowed hard in the boat until four o'clock in the afternoon; and then have worked at the windlass in cutting in the whale until three o'clock the next morning.

Then I have gone to work at trying out the oil, and for eighteen days in succession have worked constantly, with the allowance of five hours only out of the twenty-four for rest. I would frequently be so weary that I would sleep while standing on my feet: and I have had more than one good nap at the mast-head, and even at the helm in calm weather.

I have seen the time in which I could sleep as sweetly beneath a beating rain, as any landsman does on his downy bed; even though when I awoke and arose, the water would run in streams from my wet garments.

I was in the habit, for some time, of sleeping at the top-gallant mast-head; but, upon a certain occasion, our cooper, Charles Tucker,‡ a man that so much loved money, that he tried to cheat his shipmates in every way he could, saw me nodding at the lookout pinnacle, and complained to the mate. He was afraid I should fail to see some whale.

The mate cried out to me to know what I was about.

I answered him, "not much of anything; only dreaming of home." He asked me then, what I came a whaling for?

I replied, that really I did not know, but I was sure it was not for money.

"Well, well," cried he, "you are a whaling now and so long as you are on board a whale ship, you must do as whalemen do."

So soon as a whale is killed, a signal is made from

---

‡Author of the journal of which excerpts are printed as Appendix III.

the boat to the ship. She draws near, and the main yard is hauled aback, to stop her headway. The whale having been brought alongside, is made fast by a chain.

A cutting fall is suspended from the main masthead, and let down to the deck with a fish hook attached to the end of it, which weighs about two hundred and fifty pounds. This hook having been inserted into the exterior flesh of the whale, called the blubber, the work of reeling off the surface of this mighty cocoon commences. The blubber from which the oil not found in the head of the whale, is taken, surrounds the whole body, resembling a skin about one foot in thickness. After the hook has obtained a fast hold, the fall is taken around the windlass, and as it is drawn up, one man on each side of the hook, cuts the blubber with a long sharp spade, made as keen as a razor. In this operation the whale continues to revolve, until the whole exterior has been removed. When the strip of blubber reaches the top of the fall, it is cut off near the deck with a long boarding knife, and lowered into the blubber hold of the vessel. After this, three or four of the hands are obliged to jump into the hold and cut it up into "horse pieces," each being about two feet long, and a half of a foot thick. In performing this duty, the men get completely covered with oil from head to foot.§ Every time the ship rolls heavily, if the workmen are not

§Tucker describes all hands as "up to there eyes in fat."

Three whaling scenes. Steel engraving, from *Iconographic Encyclopaedia*, 1851. (Courtesy Kendall Whaling Museum)

very careful, they will slip and fall, face and eyes, into the oil and blood.

Some portion of flesh will be attached to the blubber, which soon putrifies: and the consequence is, that after working a few hours in the blubber hold, a sailor will acquire a very disagreeable scent.

Few can imagine how our crew appeared, when every one's person was covered with oil from head to foot, and every face was blackened by the smoke of the furnaces employed in trying blubber, and with iron rust, from the hoops of our casks. Even our decks were often covered with gory and black skin, so that it was difficult for any one to stand.

During the first part of our cruize, all of the crew cherished a dislike to me, on account of my being "a gentleman's son," and sought every opportunity of throwing reproach upon me, and of keeping me occupied in the most disagreeable labours.

One day the mate asked me what I came to sea for, and I replied that it was to see how sailors lived. At this he was offended, and took good care to give me more knowledge than I desired. I saw how sailors lived to my heart's content. I was the smallest person on board the ship, and to gratify his spleen in punishing me, he stationed me in the blubber hold, to pitch up the pieces on deck. Many of them were larger round than my waist, and weighed nearly as much as my whole body. The work was very tiresome, and sometimes I would stagger and fall on the slippery mass, under the weight on my fork, and be nearly immersed in oil.

But I was determined that the mate should not be able to triumph over me, by knowing how badly I felt. Frequently he would come to the hatchway for the purpose of tantalizing me, and ask if I had found out how sailors lived. In bravado I would tell him, that "I had a first rate time down there," and that so soon as I could get my work a little ahead of the receivers on deck, I could sleep on a board on the blubber.

When the "horse pieces" had been pitched on deck, they were carried to the "mincing horse," a kind of bench on which they were cut into slices, of the size of the hand, and thrown into large pots kept hot by the burning of the scraps, or cracklings, under them. From these pots the oil was frequently bailed out, and deposited in a cooler, whence it was run off into casks on deck: and again, when sufficiently cool, by means of a tub and hose, it was conveyed into casks in the hold of the ship.

So soon as a whale has been tried, the decks, rail, masts and rigging have to be washed off with ley and soap.

During the voyage I became so much disgusted with the conversation of the forecastle, that I was glad to accept of the berth of steward which was offered me. I anticipated that I should find a little more quiet and decency in the cabin than I had elsewhere experienced, but I soon found, to use a vulgar expression, that I had "jumped out of the frying pan into the fire," for the captain was the greatest blackguard in the vessel, and some of his officers were but little better than himself.||

It humbled my pride not a little to wash dishes and make bread, and it was not long before I desired again to be a sailor bold and free, amid the tar and slush, in preference to a lackey amidst flour and dish-water. I learned, however, many useful accomplishments in the line of housewifery; such as that of making bread, setting the table, cleaning knives and forks, and preparing pies out of whale's meat and molasses.

Bread and meat cooked in sperm oil, taken out of the trying pots, are, without equivocation, very fine food.

Having acted as steward about two months I concluded that I was sufficiently accomplished in culinary matters; for I was an apt scholar both in forming and demolishing pies: so I solicited of the captain leave to resign my official station.

I was tired of the business; and I knew also that several of the boat-steerers were dissatisfied with me; for they sometimes stole pies, and disarranged my pantry, on which account some quarrel had existed between us.

The captain agreed that if he could persuade

||In his autobiography (see Appendix I) Ely comments on "the unmentionable obscenity, vulgarity and degradation that rendered a whole ship a floating hell. Even the names of mother and sister were not too sacred to protect them from obscene language, and though I was terribly profane myself I heard blasphemy in the forecastle and from the lips of the officers that made me shudder. . . ."

Seth Cole to take my place I might go into the forecastle again. Cole having been called into cabin, promised the captain that he would take my berth; which he was willing to do from his intimacy with the commander, and from his promise to teach him all he knew about navigation.

I was not present at their negotiations; but I found he had determined not to enter the cabin empty-handed. Coming from the captain he told me, with a very serious face, most positively, that he would not relieve me from my office, unless I would give him twenty-five dollars and two pair of pantaloons.

The cook had told me but a few moments before this proposition was made, that Cole had agreed to accept of my place; but with this knowledge I could not out-yankee him, and was obliged to give him an order for the amount on the proceeds of my voyage.

On leaving the cabin I forgot to remove with me my journal and a number of private letters, which I had prepared to send home by the first ship which we might meet. These contained an account of my voyage, and expressed my opinion of the ship's crew in no very guarded terms. The captain I had represented as a mean puppy; and expressed my regret at having become steward, because the inmates of the cabin were worse than those of the forecastle.

The Captain got possession of these writings, to my future annoyance; for the cooper in a dispute, threw up to me some expressions of private matters which the captain had read to him. This reading of my private papers at first excited my indignation; but I finally consoled myself with knowing that the hands aft had perused my dissertation on their infamous character. They knew, without my uttering to them any mutinous speech how much I abhorred them; and the only way in which they could retaliate, was by saying frequently "some one has jumped out of the frying-pan into the fire."

Among the many good attributes which our captain claimed, was that of being a temperance man, whenever he was called upon to treat his hands with grog. During the whole voyage he sent three junk bottles forward, with the contents of which he professed to medicate fifteen men. It was well for them that he gave them no larger amount of ardent spirits.

When I signed the shipping list, it was proclaimed that the Emigrant was a temperance ship,# but I found before we left the wharf that there was a barrel of gin on board; and before our return this barrel was empty. The captain kept it wholly in his charge, and if we could credit his word he never

---

#The official shipping list for Ely's voyage on the *Emigrant* is not available, but that for the immediately following (February 17, 1847) merchant voyage to Cuba supports his contention that the vessel was ostensibly a temperance ship. No spirits or sheath knives were to be allowed on board. Again the agreement with the crew for the voyage of August 7, 1847, stipulates no profane language and no grog. These documents are in the collections of the Rhode Island Historical Society. Though it lists tea and coffee, the List of Provisions for the *Emigrant* preserved in Lincoln's journal does not mention liquors.

drank a drop of it. But if men can neither doubt their own existence, nor the testimony of their five senses, I must know that our captain did drink intoxicating beverages; for I have seen him drink them, and have known him to be under their unhappy influence, even while he solemnly declared that he did not know the taste of such liquor.

He may, however, be a philosopher, and hold that our perceptions are fallacious.

One morning, when we were trying out a whale, he came on deck, and feeling both cross and strong, either because his supper had not agreed with him, or from some other reason, he began to grumble about the cutting of the blubber, that it was not fine enough. It chanced that I was standing at the mincing horse, at the time, and holding on to the blubber with a hook. The mincer threw the blame of its not being cut into sufficiently small pieces on me. He cursed me and said, "d—n the puppy; if he don't hold on, knock him in the head!"

This roused my ire, so that I caught up a piece of blubber with my hook, and dashed it with such force on the mincing horse, that the oil flew all over the captain and the deck.

Stretching himself back he aimed a pitchfork at my breast, and damning my soul said, "dont show any of your spite here, or I will heave this fork through your heart." I stopped and looked him full in the face with my breast bared, and he cowered down; and for a few mornings afterwards was as pleasant as May. He saw that I was not afraid of him.

The Captains' Temperance and Tyranny.

Most generally, the officers of whale ships take every opportunity of running over any man who fears them; but when they find one resolute and undaunted, they have some respect to their own safety. Necessary self defence must be determined on in relation to savage men, or you will have no security against their oppression.

For twenty-seven months I scarcely heard one kind word, and on my return to land, an affectionate expression, from any friend, was sufficient to produce tears of joy.

Tristan da Cunha. The *Emigrant* raised this rugged island March 31, 1845. Lithograph, unidentified. (Courtesy Kendall Whaling Museum)

CHAPTER VII.

*Isle of Tristan D'Acunha.—
Sea-Fowl and Eggs.—Sea-Elephant.—
Sunfish.—Gales.—Isle of Bourbon.—
Isle of France.—Tombs of Paul
and Virginia.—Beautiful Cemetery.*

ON the thirty-first day of March, 1845, we made the Island of *Tristan D'Acunha;* one of the most grand and wild in appearance which I have ever seen. It rises 8326 feet* above the level of the ocean, and hides its hoary head in rugged grandeur above the clouds. It seems to vie with the tower of Babel in its efforts to overlook the battlements of heaven, and the throne of God. Yes, there it stands, in defiance of the

*This and other facts about the island Ely has taken from some contemporary account. Ely's figure of 8326 feet is exactly the same as that given in M. Matte-Brun, *A System of Universal Geography*, ed. James G. Percival (Boston, 1834), II, 149.

tempest's breath and ocean's rage, to fill every beholder with wonder and awe. It was peopled by old Governor Glass, an American, and a few other families, almost as rough as their mountain sides. The inhabitants are about one hundred and fifty. All the adults were said to be married, except one daughter of the Governor. It was also reported and believed, that the Governor has offered any American a house and one hundred acres of land, who will marry her and live on the island. A fine opportunity this, for any young man who wishes to play Robinson Crusoe!

The first settlement is said to have originated in the wars between England and France. The British Government placed a small body of troops on that island to prevent the French from taking possession of it. After the return of peace some few soldiers remained there, and sent to the cape of Good Hope for a cargo of wives.

They have very fine Irish potatoes, Indian corn, pumpkins, and cattle; besides an abundance of wild goats, deer, sea-elephants and albatross.

Some of the elevated portions of the island abound with the eggs of every kind of sea-fowl, which can be obtained for food at any time. Formerly persons who visited the place to obtain the oil of the sea-elephant erected a tent on shore and sustained themselves by eggs. Frequently they broke all the eggs in a roost, that they might get fresh ones next day.

The tender affection which the young birds of the island discover for their parents is remarkable.

When the old birds become infirm, from age or injury, the young ones supply them with the best fruits of their industry, and you will frequently see around their resting places, fish which has been deposited for them by their duteous offspring.

The birds of many kinds were so numerous on a desolate island which we visited, and so little afraid of man, that sometimes we could scarcely avoid treading on them; and found it difficult to drive them out of our way.

One of the most singular inhabitants of the deep is the sea-elephant. This animal is only to be seen during six months in the year, at the expiration of which time it swallows a large number of stones and sinks to the bottom of the sea, or as near it as is practicable.

Having obtained a good supply of fresh vegetables, we sailed from Tristan D'Acunha for the cape of Good Hope, with a fine double reefed topsail breeze, and for forty-eight hours averaged eleven knots an hour.

A few degrees west of the Cape we took a large sunfish, which required all hands and a four fold tackle to hoist him on deck.† We found from twenty to thirty good sized fish in him, besides a large quantity of squid.‡

---

†February 2, 1845. See entries for that date in the Tucker and Lincoln journals.

‡SQUID is the food of *sperm* whale, which is a fish resembling a jelly; which will sometimes cover an acre.—BRIT is the food of the *right* whale. [Ely's note.]

His gills were about three feet in diameter, and of a most beautiful white, much resembling a ruffled shirt bosom.

Before reaching the cape we had a very heavy squall of wind accompanied with hail.

At the time it struck us, I was below in my bunk. At the call of all hands, I sprang on deck in my thin flannel drawers; and in going out on the topsail yard the sail blew out, and knocked me off the yard. In falling I just caught by the stirrup with one hand, and so escaped the billows. In leaning over the yard to take in sail, the hail pelted me in the back, and through my drawers, so that to use a sailor's expression, I hopped about like a monkey in the lee back stays.

Not long after, off the Magullas banks,§ near the Cape of Good Hope, we encountered a most fearful gale;|| in a place which is called "the grave-yard of the seas;" for there is scarcely a mermaid's nook beneath those waters which has not been decked with the spoils of many a gallant ship. A short time previous to our arrival in the place, five fine large East Indiamen had sunk beneath the waves.

This gale was the only one in which the sea came up to my utmost conceptions of it prior to my voyage.

---

§Agulhas Banks.

||See the entries for April 14, 1845, in Lincoln and Tucker. According to the former, in addition to the two boats and gear they also lost a hog, who "probably . . . went to look after the boats." One hopes that it was not the cook's pet pig.

Indeed a man might go to sea for thirty years, and never realize his idea of the might and majesty of the ocean, except in some peculiar spot, and in unusual circumstances. Our officers who had followed the sea for some years, all remarked that they had never seen the waves rise so fearfully high as they did in this instance.

The officers felt alarmed for the safety of the ship; but I, like an infatuated child, gazed for hours in pleasing wonder and admiration of the billows which rolled over us.

When I first came on deck in the morning I looked over the stern, and there, as high as our top-gallant, was a black mountain of water. I could just see a speck of white foam on its brow: but as it rolled on toward us, gathering strength as it advanced, it seemed to leap into a mighty cataract, sweeping over our decks, and roaring around us with the noise of a thousand lions. The seas were so heavy, and the wind blew so furiously, that we had no alternative, but were compelled to run before them, under a close reefed main topsail, and foresail, to keep ahead of them, even though we expected every moment to see the sails fly into rags; for had the ship broached to, or had one of the seas broken over our taffrail, we must inevitably have had a coral pillow, down far, far in the unfathomable Hall of Neptune,

> Where many a sailor, bold and free,
> Now slumbers beneath the dark blue sea.#

#Unidentified.

It blew so hard, when we were reefing the top-sail, that one of our men was obliged to take hold of the rigging and tread down the sail; in doing which a puff of wind blew him strait out like a ribbon. He held on by his hands until some of his shipmates caught hold of him and drew him on to the yard. In this gale we lost our two quarter boats, stove our gangway and quarter-rail, and washed away our binnacle. During a part of the time, the water on deck was up to my waist; and sometimes even over my head, when standing by the mainmast.

At the time the sea struck us which carried away our waist gangway, the mate told me to make fast a rope across the breach, to prevent the men from being washed overboard. While I was engaged in doing this, two seas came, one immediately after the other, and swept me completely under, so that for some time I did not know whether I was on board or not. I clung however to the rope, and after a few moments found myself hard against the lee rail.

It may seem boastful to say it, but nevertheless it is true, that I was never terrified but once during my whole voyage. Although I was often in imminent danger, yet I always had a strong desire that the wind might blow harder; and whenever I perceived the gale to abate, it was with regret.*

---

*In his autobiography Ely attributes his lack of fear to his mother's dying blessing: "Though in after life, for a time, I became a wanderer from my earthly home and my Heavenly Father, my mother's dying blessing never left me, and when in the midst of a terrible storm at sea,

Indeed I gained the hatred of many of my shipmates because I said that I loved a storm, and always sung during the continuance of one. They used to call me a Jonah, and tell me that I was crazy; and I do not much wonder that they suspected me of being *non compos mentis:* for I always felt a strong desire to sing some wild and loud song in a tempest. I could realize the feeling of the Old Norse boatman, spoken of by Sir Walter Scott,† whose voice grew with the storm.

I have frequently stood looking to the windward after we had taken in sail for a coming gale, and watched the dark banks of clouds, as piled one upon another they rose amid the glare and crash of Heaven's artillery: and as the wind moved in fitful gusts through the rigging, I longed to sing some low and melancholy strain. Then, as the gale approached, and the waves began to curl their shaggy mane, and the increasing wind would shriek through the cordage, my song would grow louder, until, when the storm was highest, I would find myself singing at the top of my voice some impromptu suggested by the occasion.

Once, and only once during my voyage did I feel serious alarm. It was during a very dark night.

---

when I stood lashed to the mast as the huge waves swept over the ship carrying all before them, the thought of her prayer banished my fear, for I believed that my mother's prayer would be answered." Even in his yachting days, however, he had taken "more pleasure in strong winds and rough seas than in a calm."

†Possibly a reference to Sweyn Erickson in *The Pirate.*

We had all her sail on the ship, and were standing along under a moderate breeze, when a sudden squall laid her over almost on her beam ends, and made every stay and brace in her crack. We clewed up and down everything as soon as possible; and the mate sent me aloft to furl the fore-top-gallant sail. If ever a prayer was sent up to heaven in short order, from a sense of danger, it was when I stepped from the shrouds on to the yard; for I had not the least expectation that I should ever step off from it again.

The lightning appeared to play on the ropes on which I was standing, and all over the ship; and the masts quivered like an aspen leaf.

Death seemed to be on one side and on the other, and I thought it best to die in the performance of my duty.

It happened that the only coward on board was sent up with me. For some time I could not persuade him to get on to the yard. He started and went part of the way down the shrouds three times. I sung out to the mate to send me a man up, for the fellow he had sent was such a coward that he could do nothing. This roused his pride so much that at length he got on the yard and assisted me. It was a long time, however, before we could make any impression on the sail; for when clewed up, the wind would make it stand straight out, and resist all indentation from the blow of the fist.

During the whole voyage we had an unusual amount of bad weather, and for six months of the time, not sixty days found me with a dry stitch on my back.

A short time after leaving the Cape,‡ we experienced another gale which carried away our fore-top mast, the main top-gallant mast, and the jib-boom. The ship was a dismal looking wreck, with her masts hanging over the side, and her rigging strewed like dishevelled tresses upon the deck and rail.

On account of these disasters we aimed for the Isle of France; but our chronometer being somewhat erroneous we made the Isle of Bourbon,§ supposing it to be the former, and sailed all around it, seeking for the port. This is one of the most beautiful islands of which it is possible to conceive.

In the back ground you may see the high volcano rising in gloomy grandeur above the clouds. It seems to guard the lovely scene below, like some swarthy browed warrior watching over a slumbering infant. In front and beneath lies a beautiful and gently sloping plain, dotted here and there by a golden rice field, a green orange grove, or a white cottage, with the blue sea rolling up to its very base, as pure and ether-like as in mid ocean, except where here and there some gallant ship, or graceful pleasure barge, spreads its white sails in the sun-light. It is just such an island as we must suppose Moore had in view, when he wrote those beautiful lines,

---

‡May 6, 1845. See Lincoln and Tucker.
§May 14, 1845.

Hindu costumes as Ely saw them on Mauritius. Lithograph from a series commemorating Perry's voyage to Japan, 1852—1853. (Courtesy Mrs. Herbert DuBois)

Isle of France.

"Oh! had we some bright little isle of our own,
In a blue summer ocean, far off, and alone."||

The climate of the Isle of Bourbon is most delightful. The principal inhabitants are French people and creoles descended from them. Their principal occupation is that of raising sugar.

Having discovered our mistake we sailed for the Isle of France, and arrived there the next day. Near the port we were becalmed, and a steamer towed us in.#

The immense quantity of shipping usually found here is truly surprising. There is probably no island in the world in which you will see the flags of so many different nations displayed as in this. Indeed if a traveller is desirous of seeing the manners and customs of all the tribes of men at once, let him land at the Isle of France. There he will find the turbaned Turk in his mosque; the Arab with his bald pate; the Chinaman eating with his rice sticks; the Hollander and German with their pipes; the sandy haired Scotchman with his bagpipes; the Englishman and his brown-stout; the Irishman and his whiskey; the Frenchman with his love of frogs, dancing and military glory; the American with his readiness to trade in all sorts of notions; Jews with their golden trinkets, and money to loan on pawns; Lascars, Malabars, Hindoos, Ceylonese, Burmese, Portuguese, Spaniards, and woolly headed negroes, all mixing together, and

||The first two lines of a poem in Moore's *Irish Melodies*, the first line being the title.

#May 17, 1845.

each living after his own fashion and religion.

The Malabars in the island are a very handsome race of people; erect in their persons, graceful in their carriage, with fine, open features, and clear though dark complexion.

There is on the island one of the most beautiful cemeteries which the world can boast; and one in which the tender regard evinced for deceased relatives is truly gratifying. It is the custom of the place early every morning, to place garlands of flowers, and boquets upon the graves. There is a large cross in the midst of the grounds which is regarded as an altar on which the people offer the choicest flowers and plants of the tropical regions. It is often loaded with them. Many of the plants of the island are exceedingly beautiful. I have seen the cactus used as a hedge.

Whilst on this island I visited the tombs of Paul and Virginia, the history of whose life, though commonly regarded as a fable, is nevertheless substantially true.*

Their tombs are situated in the same forest in which they were lost, and which is now owned by an English gentleman.

You enter these grounds by a beautiful shaded avenue, ornamented on each side by rich beds of flowers, which leads to the front of the proprietor's house. Walking around this, you come upon a placid little lake, resting beneath the shade of the over-

*A reference to the sentimental romance *Paul et Virginie*, by Bernardin de Saint-Pierre.

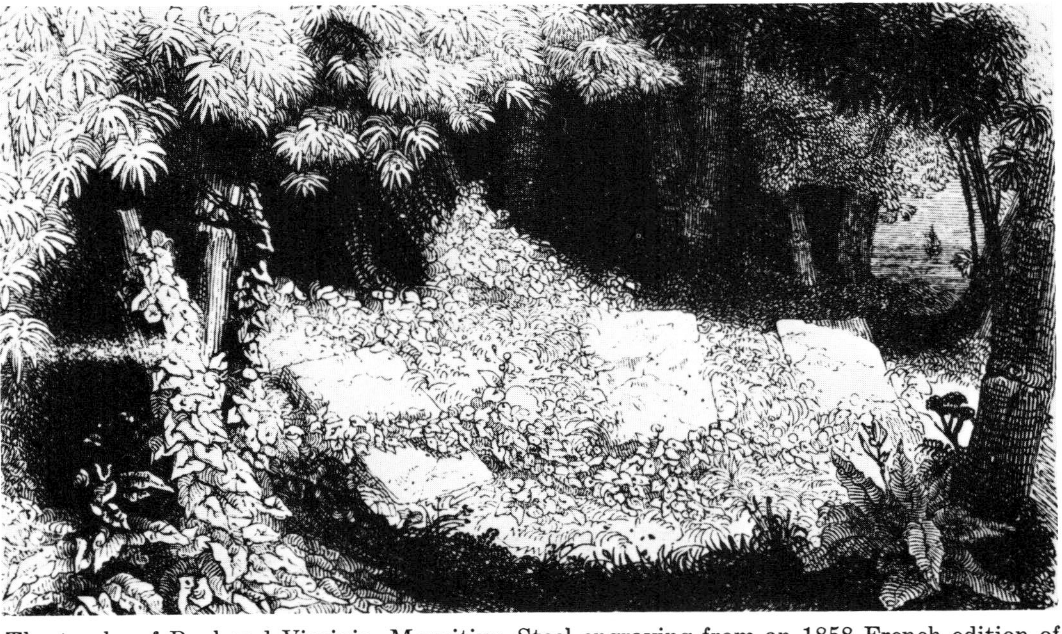

The tombs of Paul and Virginia, Mauritius. Steel engraving from an 1858 French edition of Bernardin de Saint-Pierre's romance. (Courtesy Olin Library, Wesleyan University)

hanging forest. On the right side of this lake is the tomb of Paul, and on the left that of Virginia.

When I approached the mansion, I was dressed in seaman's attire, and of a matron who was seated on the piazza I asked leave to visit the tombs. She granted it, and called her grandson to accompany me and show the way. He was a bright haired little fellow, who took my hand and told me he knew I was a sailor, because my breast was bare.

We walked around the tombs together, and he picked up a fragment and gave it to me, saying, "grandma told me not to break off any, but here is a piece already broken off, which you may have." I took his offering, and placed it in my bosom; when he said very archly with a smile, "you are putting Virginia next to your bosom are you? I know that sailors always put things in their bosom; for when I came from England I often saw them drop their pipes and tobacco out, when they went aloft to reef the topsails."

We remained in this port for six weeks,† and each watch had every other day on shore.

After having seen the curiosities of the island, there was little pleasure in being among the inhabitants, for they all look down upon mariners as a low and degraded set of beings; as indeed very many of them in former times were; and poor Jack Tar could find no companionship except in some tavern or grog-shop.

---

†Actually three weeks: May 17 through June 5, 1845.

CHAPTER VIII.

*Hard lot of Sailors.—Some Deserters.—
Mickey's rebellion—Sperm Whales—Right
Whales—Whalebone and Brit.—
Nossi-Be—La Pont—Madagascar—
Its Inhabitants.—Customs, &c.*

WHEN sailors have been shut up within the prison walls of a ship for months, and sometimes for years, can any wonder that they crave society, and are a little extravagant, so soon as they touch their mother earth? They wish to see and converse with some other beings than the crew of their own vessel. They must have some place of resort; and where decent sailor boarding-houses cannot be found, they are compelled to abide in such as they can find. In foreign ports, especially, a tar is avoided and gazed upon as an ass or a lion; and because people expect no good of him, and show him no civility, he is often reckless in his conduct. "Nobody cares for Old Joe; and therefore Old Joe cares for nobody." He is like Job, when his relatives and former friends refused to know him; or like the good man in the Psalms, who said, "no man careth for my soul!"

It is a great error to judge that all sailors are ignorant; for many of them have received a good common, and some of them a classical education. Many of them are well informed in matters of geography, customs, manners and commerce; and were they treated as men, and allowed the ordinary privileges of freemen, they would soon show themselves worthy of a reputable standing in the community. At present they are exposed to be treated like dogs or slaves at sea, and as outcasts on shore. Let the benevolent bestow their benefits on seamen, treat them kindly, teach them the doctrines and duties of religion, seek to reform them, afford them moral and intellectual entertainment in port, and furnish them with the Bible and other good books, and our ships would soon do more for the conversion of the heathen, than a dozen cargoes of missionaries.

During our stay at Port Louis, six of our hands ran off, and several more would have done so, but lacked opportunity. To supply the places of those who had absconded, we shipped two men who had some time before deserted from a New Bedford ship. These poor fellows had suffered much, for after leaving their own vessel, they were obliged to live in the mountains on roots and wild fruits.—Then they were imprisoned for six weeks, and were compelled to sustain life with rice, and to sleep upon the ground. After their incarceration they worked their passage from the Sea Shell Islands‡ on board a latteen to the Isle of France. Here they could not obtain work because they were Americans, so that for want of money and other means they slept under the market-house, and ate raw stock fish for sustenance. Occasionally some foreign sailors would give them a few crumbs of bread. One of these men, Barnard,§ was a very fine fellow, and both of them were most heartily sick of whaling; but they had no better way of returning to America than by shipping on board our bark. The night before we left the Isle of France, all hands, according to a long established, miserable custom, "got royally drunken" under the false pretext of banishing gloomy thoughts at re-embarkation and leaving the habitations of civilized man.

During this drinking spree, Mickey, the Irishman, determined to go on shore, and went aft to ask permission of the second mate, after having been refused by the chief mate. Mickey being "three sheets in the wind," told him, that if he did not allow him to take the boat he would jump overboard. The noise made by him brought all hands on deck, who stood watching for the issue of the affair.—The mate persisted in withholding permission; and Mickey, as

‡The Seychelles Islands, six hundred miles northeast of Madagascar.

§Bernard McClusky. According to the Crew List, four men were signed on at Port Louis.

good as his word, jumped overboard and began to swim for shore. A boat was immediately lowered after him, and while the men on deck shouted, "run, Mick, run, run," all the officers kept crying out, "stop him! stop him!" He reached the shore; but unfortunately for him, a sailor near him, where he landed, thought the cry was, "stop thief! stop thief!" So he caught the runaway and held him until the boat's crew landed and took him a captive back to the Emigrant.

Being once more under the dominion of the Captain, orders were given that he should be placed in irons. To execute the order was the difficulty, for Mickey fought desperately and bravely all who approached him, and none of the hands would aid the officers except Old Joe, who was so much intoxicated, that he did not know what he was about.

For this co-operation he was teased during all the remainder of the voyage. He had at the time that inspiration of grandeur which much grog is apt to impart; for he thought himself the commander, and said to me a few moments after the irons were on, that he was sorry for Mick; that he had acted very much like a fool; for had he only asked HIM, he might have gone on shore. Rum will make any tar a captain, without the power of office, for a little time.

The next morning we hoisted topsails, and with canvass filled with a favourable breeze, and with the hearty cheers of our countrymen as we passed their ships at anchor in the bay, our crew

Sperm Whales—Right Wales

"Vela dabant laeti, et spumas salis
aere ruebant."||

And we might say with Byron,

"Once more upon the waters,
Yet once more,
And the waves bound beneath us,
As a steed who knows his rider."#

A few weeks after leaving Port Louis, we took our first SPERM WHALE. This species of fish is captured in the same manner as the RIGHT WHALE.

Right whales are, in general, much larger than sperm whales, and very different in form, having no hump on their back and possessing much more extended viscera. The head of the right whale is commonly covered with barnacles, or a species of small oyster; and with whale lice, or small crabs; so that the head under water has the appearance of a rock. The resemblance is so great that when we obtained our first whale, which was of this species, one of our green hands cried out in alarm to the captain, "look out sir! you will be on that rock!" He soon found, however, that the rock had life, and could give vent to its agony when the harpoon entered it, in a mighty roar which could be heard above the roar of the ocean, for some miles around.

||Vergil, *Aeneid,* I, 35.

#Byron, *Childe Harold's Pilgrimage,* Canto III, stanza ii:

Once more upon the waters! yet once more!
And waves bound beneath me as a steed
That knows his rider.

A whaleboat attacking a right whale, with a whaleship cutting-in and boiling in the background. Color lithograph by Currier & Ives. (Courtesy Kendall Whaling Museum)

Whalebone and Brit.

The black bone which has been so much used by ladies, is taken from the mouth of the Right Whale, in which it forms an arched roof; being laid in slabs which come near together at the top, and are placed at a short distance apart; like Venitian blinds. These slabs of bone are fringed with long hair in the interior of the mouth, for the purpose of catching the BRIT on which he lives.

This BRIT is a reddish substance, which rises at certain seasons of the year, and floats upon the surface of the water. It very much resembles Indian meal, and covers not merely acres, but at times square miles of surface. It appears to be a conglomeration of unnumbered millions of piscine animalculae.

The right whale scoops up this substance with his enormous lips, and pressing it through the slabs of bone in his mouth, catches it in the heavy fringe, and licks it off with his enormous tongue, which will, of itself, make from ten to forty barrels of oil.

On the outside of the blubber, or fat of the whale lies a coat of fine short hair, moistened and matted together by a dark glutinous substance.

This coat of matted hair is covered by several integuments of black skin, which much resembles india-rubber; is very porous, and when cut or scratched, emits profusely a lubricating substance, which being neither oil nor glue, for want of more science, I must call *slime*.

The right whale has a forked spout, which he throws on each side of him, from the spout hole in the back part of his head.

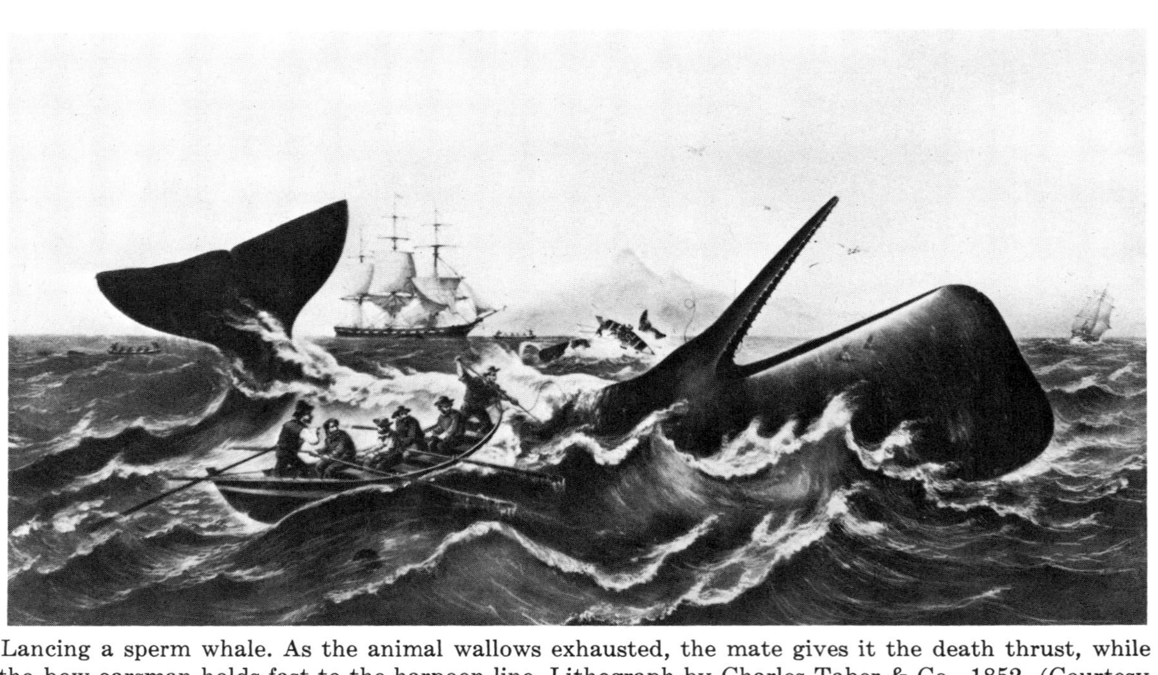

Lancing a sperm whale. As the animal wallows exhausted, the mate gives it the death thrust, while the bow oarsman holds fast to the harpoon line. Lithograph by Charles Taber & Co., 1852. (Courtesy Kendall Whaling Museum)

A *sperm whale* has a large oblong square head, shaped like a brick-bat, of polished black, out of which the sperm oil is taken; with one large and several small humps on his back. His jaw resembles a harrow, and is supplied with large white teeth. His spout hole is on the tip of his nose. In other respects he is much like a right whale.

After taking our first sperm whale we experienced another severe gale,* in which we were in imminent danger of being stranded, but finally escaped in safety.

On the 27th of June, 1845, we cast anchor in Nossi-Be, a small island in Passandava Bay, on the north-west side of Madagascar. On this island is a French and Arab settlement. It is retained by France as a rendezvous for her men-of-war. The French town, La Pont,† is beautifully situated on a small prairie, surrounded with lofty hills, whose fertile sides are redolent with many a bright exotic, which breathes its fragrance forth upon the bay, to gladden the sailor's heart; or as imagination would bid us dream, to supply the mermaids of the Mozambique with odour for their hair, as sitting beneath the shady cliffs they comb their tresses o'er the moonlit wave.

On every side this lovely bay is hemmed in by mountains, clad in verdant robes; verdant did I say? Yes, verdant, except where here and there some

---

*On June 22, 1845, Tucker reports " a continued gale the last 7 days." "The old ship," he says, is "a jumping and pitching . . . like a sick porpoise."

†Probably the modern Hellville.

golden rice field, in lieu of breast-pin, decks the bosom of the hill, or where the blue smoke of the native hut is seen to arise in sun-lit air, or wreathe a crown about the mountain brow.

The principal productions of this island are bananas, plantains, lemons, oranges, pine-apples, melons, sweet potatoes, and rice. This last article grows luxuriantly on the hill side, and not, as is commonly the case in other places, on low marshy ground.

Madagascar is inhabited chiefly by several tribes of savages; the most numerous and powerful of whom are the HOVAS. They are a race of light coloured people, for a tropical climate, with black eyes, and long black hair. Their forms are generally fine, and their movements graceful. Some of the females are beautiful, but most of them are exceedingly affected in their airs; lascivious, and well instructed in those thousand little arts by which the power of beauty chains the soul. Their costume is simple, graceful, and picturesque; consisting of a flowing robe thrown over one shoulder, and descending sufficiently low to seem to hide, yet set off the more, the beautifully turned ankle, and arching instep. One finely moulded arm, shoulder, and budding breast, in Circassian style, is left uncovered.

They speak a musical and soft language which seems to be compounded of native, French, Spanish, and English terms. Most of them are melodious singers. Frequently I laid myself down on the bow of our vessel, as the sun was sinking in the West, and listened

The coast of Madagascar, with a vessel offshore. A nineteenth-century German lithograph. (Courtesy Kendall Whaling Museum)

to the chorus of their voices, as it came wafted on the wings of zephyrs over the stilly sea.

Towards evening you will generally find them seated in circles before their doors, with the best singer in the centre, carrying on the burden of the song, while those around her join in the chorus.

The inhabitants of Madagascar come to maturity at a very early age. I think I have seen a mother and daughter, neither of whom was over ten years old.

It is considered a decided mark of worth and morality in a boy of twelve or thirteen to be willing to marry.

Marriage does not continue during the lives of the parties, but only for a stipulated term; which is generally one year; and at the expiration of the term, if they mutually admire each other they are married for another period.

The practice of killing deformed infants, prevails very extensively among these islanders. They have also a custom of adopting sisters and brothers, so that no one is without them. In token and confirmation of this adoption, the parties who are to enter into the fraternal or sisterly relation, cut each other's breasts with a sharp knife, and mingle their blood in the presence of all their friends. The operation is still continued, for I saw the ceremony of affiliation practised. After mingling their blood they all become drunken, by an intoxicating liquor made of sugar-cane.

It is considered highly criminal among these people for a man to kiss his mother, or his sister. After a long absence from each other, they show their affection on meeting, by washing the feet of their relatives, and then drinking the water employed in this disagreeable service.

The women do all the work of the family and of the field; thresh out the rice, and husk the corn; while the men practise hunting and war. The last is their principal occupation, in which they use spears and muskets. The spears are generally of ebony, about seven feet long, pointed with sharp steel heads, which dexterous men will throw with great precision to a considerable distance, and kill an enemy.

Muskets are owned and used only by men of wealth and distinction.

The superstition of the people is very great, and they place implicit confidence, as very ignorant people generally do, in their prophets and astrologers; who, on their going to war, present them with charms, which they allege will prevent spears and bullets from piercing them.

They venerate also their distinguished dead, around whose graves the relatives meet once a year to hold a feast. At their first meeting on such occasions each man and woman brings a lap-full of earth, with as large a stone as he can conveniently carry, out of which materials a mound, or an altar, is erected over the place of burial. After the erection of this memorial of affection, they roast upon it one or two bullocks, at each annual festival, and conclude their rites by drinking, smoking, and dancing.

In general, they are hospitable and kind in the midst of their moral degradation. Their ideas of right and wrong, and even of decency, are very imperfect; and their notions of matrimony were such as to suit our captain and most of his crew. Our vessel, while we were in port, was a floating scene of abominations: for fathers and brothers would bring their daughters and sisters on board, and sell all the virtue they possessed for a fathom or two of calico, or a piece of iron hoop.

They look up to Americans as a superior race of beings, and strive to imitate them in every manner possible. Of course they first of all catch their vices. Hearing sailors perpetually using profane language, and more especially the officers, they deem swearing and cursing a great accomplishment; and so to distinguish themselves, and let you know that they have seen some of your countrymen before, they will respond to your salutation by a smile, and a "G—d d——n," without the least idea of the meaning of the words.

Most of the different tribes of Madagascar have their several kings and queens, for whom they profess the most profound regard. Their power is sovereign and most arbitrary. The king is often secreted in his own house, and no eyes except those of his twelve prophets are allowed to see him, on pain of death.—When the queen passes along, every one

A village on Passandava Bay, Madagascar. A mid-century French lithograph. (Courtesy Kendall Whaling Museum)

must bow his face to the ground. If any one dares to gaze on her beauty or ugliness, he is decapitated. In some tribes, the laws are not so severe as these, which obtain among the Hovas and Seclaves.

The present queen of the Hovas killed her spouse, Ladamma,‡ by poison. For a long time she has been at war with the French and the Seclaves. The French have been attempting to conquer the whole island, but there is little probability of their doing it, for the nature of the climate is such that Europeans cannot endure it. Besides, the Hovas are strongly fortified in the mountains.

It is thought by many that the English have assisted the Hovas against the French, on account of the power which the acquisition of the island would give the latter in time of war over the East India trade.

The Hovas are supposed to be the descendants of those pirates who formerly settled on Madagascar; and they certainly have some marks of civilization among them, for they build forts, manufacture powder, and clothe themselves in gay apparel. To procure the means of making powder, the people deliver up a tithe of all their oxen.

The Governor and the grandees dress with great splendor, and commonly in the richest American uniforms, bedecked with gold and silver. On public occasions they are carried in front of their troops, on

---

‡A mistake for Radama. The queen was Ranavalona I (1828–1861).

the necks of their slaves. The queen is said to have one of the most splendid brass bands in the world, consisting of more than two hundred musicians, who are kept playing before her palace.

During our cruize after we left Nossi-Be, we ran very near a desolate island called "Sandy Island," and sent a boat ashore. We found here the bows and rail of a ship, with her anchor, a bottle, and a hut; whence we inferred that some vessel had been wrecked on the spot, and that the crew had been taken off by some welcome sail.

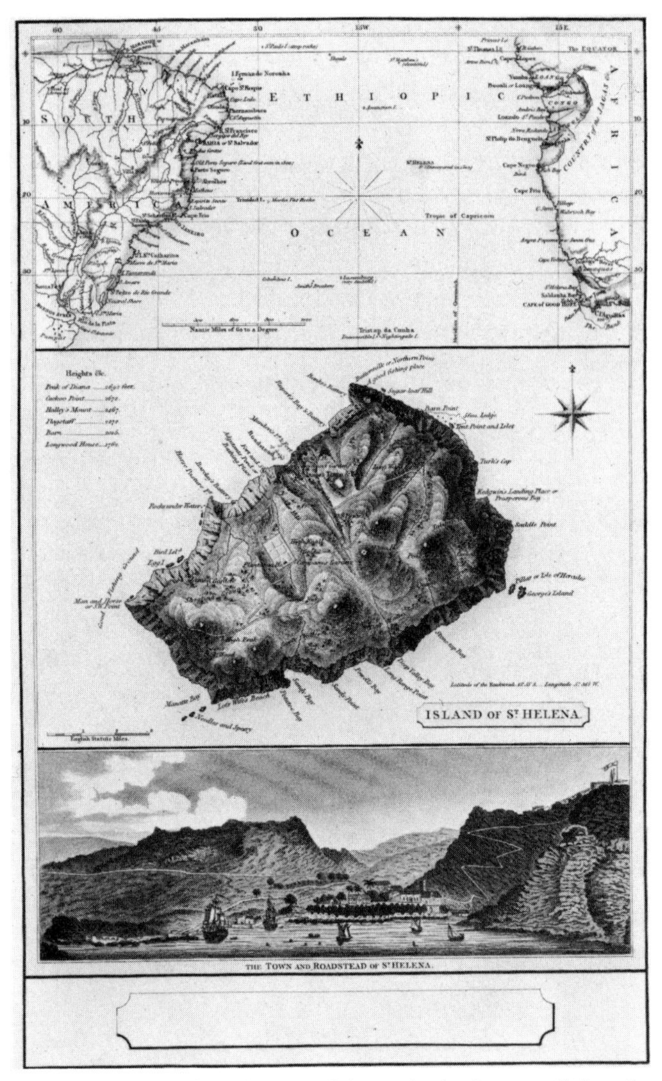

St. Helena, its location, and its principal port. Colored line engraving by Whittle and Laurie, 1816. (Courtesy Kendall Whaling Museum)

CHAPTER IX.

The Voyage home.

We cruised around Madagascar for a considerable time without meeting with any event to vary the dull monotony of our life. Our morning and afternoon watch we spent in working on our ship: our dog and night watches in spinning yarns of home, in picturing our feelings on our return home, in singing and smoking the pipe.

At length the joyful words were heard, "Square the yards for home!" and never did a whaleman's crew man the braces, and shake the reefs out more cheerily than we. There was one thing, however, which damped the ardour of our men, for many of us could not boast of more than one suit of clothing; and we had some imperfect idea of the American coast in winter. Experience soon taught us, that imagination, strong and vivid, often falls short of the reality. Our vessel, moreover, had become much worm-eaten, so that we were compelled to labour at the pumps, even in fair weather. What, then, had we to expect in a storm?

Our captain looked forward to passing the Cape of Good Hope, and approaching the coast, much as a criminal would regard his approaching trial in a case of life and death. As for myself, no danger nor hardship could for a moment check the longing desire which I had to view my native land. Death I regarded as preferable to the life I then led.

We passed the Cape, nevertheless, in safety, in a heavy gale, and bore away for St. Helena, described by the poet as the lone barren isle, where the wild roaring billow assails the stern rock, and the loud tempests rave.§ A more gloomy, desolate spot, although grand and picturesque, I never saw.

High above its foaming base, hang its dark and rugged rocks, shrouded in their mantle of clouds, which have ever clustered around the island's brow, as if to hide from heaven's sight the hero's prison. And there, far above that element which most resembled his mighty mind and changing fortunes, was the Conqueror's place of rest, until France, to do homage to his memory, removed his remains to Paris.

There were but few vessels at St. Helena on our

---

§From "The Grave of Bonaparte," attributed to Lyman Heath:

On a lone barren isle, where the wild roaring billows
  Assail the stern rock, and the loud tempests rave,
The hero lies still, while the dew-dropping willows,
  Like fond weeping mourners, lean over his grave.

arrival there,|| and we passed without much delay, to the island of Ascension, intending there to take in a supply of water,# for we were fearful that we should be driven off the American coast, and delayed in our arrival, so as to be in want of that indispensable article. We had no more water at that time, than was sufficient to last us in a direct passage home.

On reaching Ascension we were boarded by a man-of-war's boat to know if we could spare her some water, because there had been a drought in the island, which rendered it necessary for her to send for supplies to St. Helena.

The only chance, then, that remained to us, was to be supplied by catching the rain as it should fall near the line. In this, Providence favoured us. Such heavy rains fell that we caught several casks of water, of rather brackish taste. After crossing the line,* our ship sprang a-leak; so that many of us were apprehensive we should never see home. We were compelled to pump five or six thousand strokes every twenty-four hours. Pumping is one of the most trying operations of a manual kind in which a sailor ever engages. Frequently I have stood at the pumps until from

---

||They passed St. Helena, evidently without landing, December 8, 1846. Actually they saw a number of sails. See Lincoln and Tucker.

#The *Emigrant* arrived at Ascension Island December 15, 1846. As Tucker puts it, " a boat borded us from the shore and we lernt that we could not get eney watter there and consequently we did . . . not come to anchor but was O.P.H. [humorous slang for 'off']."

*About December 23, 1846.

Ascension Island, which the *Emigrant* passed in December 1846. Color lithograph, 1835. (Courtesy Kendall Whaling Museum)

exhaustion I could not see. I found too, that it affected my lungs, and made me spit blood repeatedly: but there was no way of avoiding this labour, for others disliked it as much as myself; and in general, sailors have too much generosity and ambition to allow their ship-mates to perform duty which belongs to themselves.

We did not feel the cold very sensibly until we reached the 20th degree of north latitude.

On the morning of January 17th, 1847, commenced a strong double-reefed topsail breeze from the west south-west, under which we stood on northward and westward under a heavy press of sail. About ten o'clock, A.M. we saw at a distance on the lee bow, a large black cloud rising directly against the wind. Immediately all hands were called to take in sail; but before we could accomplish any thing more than to get our sails clewed up, the hurricane struck us a-back, bearing us right over, and immersing our lee rail until the water was up to our main hatches. Its force was so great that it raised the sea right up before it in a column; and when it broke upon us, the rain fell so fast, that I really thought I should drown on the main-yard. A boy in the bustle let go of the mizzen sail sheet, and before we could haul it down, it blew into shreds. None should feel more deeply than a sailor the truth of the words,

> "Our dwelling is in the Almighty's hand;
> We come and we go at his command."†

---

†Unidentified. Apparently from a hymn or a versified version of the Bible.

For He alone it is, who can bid old Ocean's bosom heave, or lull the fearful giant into rest.

At six o'clock P.M. the wind hauled to the eastward, but still blew a gale.

> "Bermudas has not let us pass,
> And we must fear Cape Hatteras."‡

On Monday, January 18th, 1847, we had a fine breeze, and sailed with "a wet sheet and a flowing sea,"§ for my own dear native land. We were pleased to see five sail bearing our own bright colours, the proudest and fairest that float on the breeze; colours which have caught their different dies from the rainbow, and torn the azure robe of night to supply their union.

It is a singular fact that the American stars and stripes can be distinguished at a greater distance than any other national flag which sweeps over the main. These ensigns of liberty, cannot, however, defend all who sail under them; for this very day, Manuel, a Portuguese, one of the boat-steerers, without cause, began to haul about a Madagascar man, knocked him down on the deck, stripped him of his only remaining shirt, and kicked him several times in his ribs. This is the fifth or sixth time he has quarrelled with his shipmates. Once he knocked me down, when I was so

---

‡In *Two Years Before the Mast,* Chapter XXXV, Dana quotes a slightly different version of what he calls "the old couplet":

> If the Bermudas let you pass,
> You must beware of Hatteras.

§Title and first line of a poem by Allan Cunningham.

weak from sickness and hard fare, that I could not defend myself. Manuel is a powerful man, almost universally hated. The captain was evidently afraid of him, for although he had sworn "to seize up" the first man on board who should strike a blow, he allowed the offender to pass unnoticed.

On the 21st of January, we had all day another distressing gale from the north-west, with the first snow and ice which I had seen for three years. No one on shore can imagine how sailors suffer in such weather, when coming from a warm climate. My hands were so cold and stiff that I could scarcely close them when taking in sail; and the prints of our fingers were often left in blood on the frozen sails. It was necessary to carry a handspike aloft, and beat off the ice before we could furl or reef them. I had become accustomed to all kinds of hardship, but the pain in my feet and hands was such that I came near fainting, and falling from the yards into the sea, with bodily anguish. Our Madagascar hands, who had never seen ice or snow before, afforded us much amusement.

On the 23rd and 24th of January we were still harassed by gales, and it was hard to be compelled to admit, that when within two good days sail of home, we might encounter weeks of unremitting hardship, delay and danger.

On the 25th of January we passed the Gulf stream, and arrived within soundings. The water in the stream was tepid at half-past nine o'clock, and in half an hour after, when we had passed it, the sea was cold enough to make my fingers ache.

The night of January the 27th was one of bitter coldness.|| The wind whistled through our rigging, and every sea which washed over us hardened into ice on our decks and spars. Still we were obliged to stand eight hours exposed to the storm, without any shelter, and without any change of apparel when we could go to our bunks. To be within one day's sail of home, and yet not able to reach it; and to be in danger of being beaten off for weeks, if not forever, O how distressing!

On the 28th of January once more our yards were squared for home, and I stole below for a short time, to utter the praise of our gallant little ship; for indeed bravely had she borne us through many scenes of danger.

Swiftly o'er the billows' foam,
Quickly towards my native home,
Thou dost bear me on my way,
And I ask for no delay.

Fiercely on thy weather side
Beat the raging wind and tide,
And thy creaking timbers groan,
Yet thou'lt bear me to my home.

When thy snowy sails shall play,
Shadowed on Mount Hope's green bay,
Friends will bless thy noble form,
And I bless the wings of storm.

||See Tucker's vivid entry of this date.

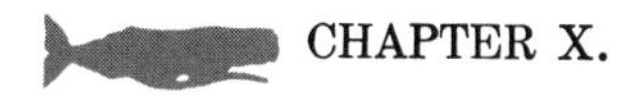 CHAPTER X. 

The Arrival and Conclusion.

How shall I begin to describe our arrival; perhaps the most joyful event of my life?

Tupper says that we should take the wheat of the idea and cast away the chaff of words:# but how shall the grain grow without the chaff? How can *ideas* be expressed without *words?*

Language is too poor to express my feelings, yet I will endeavor to make it answer my purpose.

It was a cold, clear night, the wind was from the

#Martin Farquhar Tupper, "Of Memory," *Proverbial Philosophy*, Series I:

. . . feed the mind with fatness, . . . fill thy granary with corn,
Nor clog with chaff and straw the threshing-floor of reason,
Reap the ideas, and house them well; but leave the words high stubble.

north-west, and just served to swell our sails out full to the moonlight. The crew were standing about the deck in groups; some conversing about home, some dancing, and others singing in manifestation of their joy. Every now and then some one would go aloft, hoping that he might be the first to cry out, "Land, O!" It was useless to attempt to sleep, for we were too much excited to admit of it; yet we might as well have done so, for although our imagination turned every fleecy cloud into land, and every rising star into a light on shore, yet we were not destined really to see land until day-light.

Just as the dawn, however, had fully broken,* I sprang aloft to look out, and as I reached the top-gallant mast-head, land broke upon my view right ahead; and never was more joyous and thrilling shout uttered from the bark Emigrant, than that which burst from my lips, when I sung out, "Land, O! Land, O!" Indeed I do not know but that any person less enthusiastic than myself would have considered me a candidate for the mad-house, for I swung my tarpolin and hallooed "Hail Columbia," "Yankee Doodle," "Stars and Stripes," "Home, Home," all in one breath. Then I descended to the deck, danced and sung, shook hands, and clapped my shipmates on the shoulder at such a rate, that had I been any-where else than on ship-board, I should have been taken for a Shaking Quaker, under the highest excite-ment of the spirit.

---

*January 29, 1847.

The land which we saw proved to be Block Island, toward which we stood until twelve o'clock, meridian, when the wind died almost entirely away. Here again our hopes flagged, for we saw towards night that a north-easter was coming on, and knew that if we were not able to get in before night, we should be† wrecked, for the atmosphere was dense and our vessel leaked so badly that we could not put to sea. Indeed, as it was, if we had not succeeded in taking a pilot we must inevitably have been lost: and we came near having none through the avariciousness of our Captain; who finally took the last who offered. Him he would have rejected, had he not engaged to take us in at half the usual price of pilotage.

A kind providence, which had protected us in many scenes of danger, stood by us in this last trial, and shortly after dark we came to anchor under the guns of Fort ———.‡ We had scarcely time to take in sail before it blew one of the most tremendous gales which I ever experienced. All hands, however, went below, except the two who stood anchor watch; and they, by a large fire, whiled away the time in the use of hot coffee and tobacco.

It blew so fiercely that for two days we were unable to weigh anchor. On the second day a boat's crew of us went on shore; and for the second time in twenty-seven months, I espied an American

†Emendation in editor's copy, in Ely's hand: "might be."

‡Fort Adams, according to Tucker.

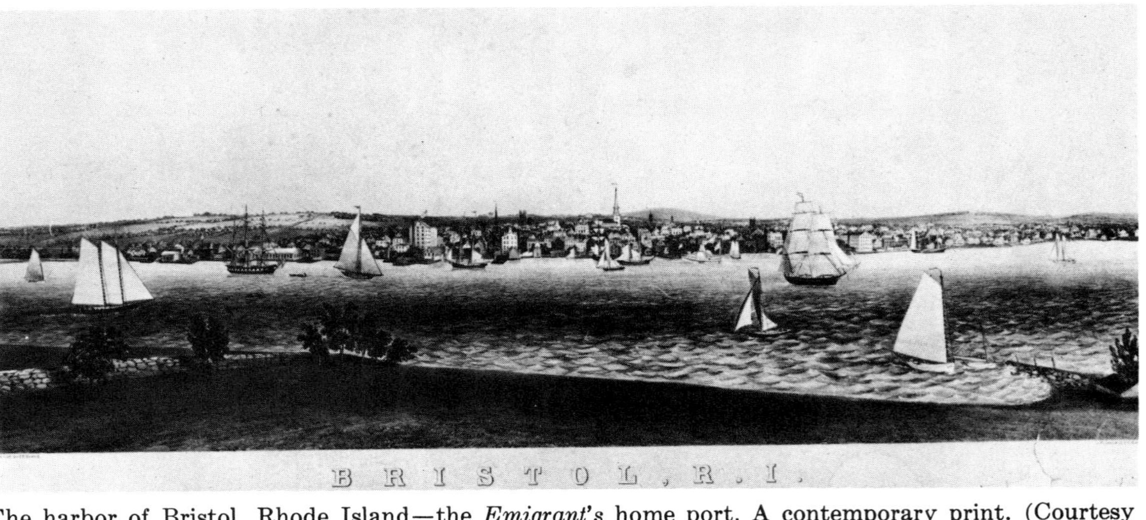

The harbor of Bristol, Rhode Island—the *Emigrant's* home port. A contemporary print. (Courtesy Rhode Island Historical Society)

woman. I darted off on a run to have a look at her, but she entered a house, before I could come up with her. My shipmates alleged that I had frightened her; and in truth, I think it must have been so, for my exterior was a rare combination of patches, tar, oil, and dirt such as her bright eyes had never before seen. If any one is desirous of seeing beauty in perfection, let him go a long voyage, and he will see it on his return. For some time all persons whom we met seemed beautiful; and our crew insisted that every man, woman and child must have been painted.

On the 1st of February, 1847, the wind having lulled considerably, we set sail up the bay, and just before dark, hauled alongside the wharf mid cheers and songs, and clewed up, and furled our canvass for the last time. For want of a conveyance I was compelled to remain in Bristol all night. I would have walked twenty miles to Greenwich, had not the snow descended rapidly; and I feared that in the darkness I might lose my way. To solace my impatience I walked the streets, roaming around in the snow; for I was too happy to sleep, or feel wet or cold. Before day-light the next morning, I took stage for Providence, and arrived there to breakfast. Finding I could have no conveyance thence until afternoon, I started on foot and accomplished my journey of twelve miles in less than two hours.

One who has never been absent from his native land, and all that he holds dear, can have but little idea of the thoughts and feelings which make the sailor boy's bosom heave and swell like the stormy ocean, when he returns to familiar scenery, and each bank and stream, each hill and dale seems to breathe the voice of "welcome." Many a sweet and tender flower of memory is called forth to bud and blossom beneath the fostering smile of Home. Thoughts, wafted on immaterial wings, float over his soul, like the breath of spring upon the earth, to warm his heart, and call forth the fragrant rose of reminiscence; whose morning dew the tear-drop well supplies. Yet the cloud of doubt and fear its influence lent to fill up the picture of this May day; for it had been months and years since I had heard one word from home, and I knew not who might be dead or living. I knew not but the silver locks of my father might then be mouldering in the tomb. Perhaps the gentle form of her who was more than sister to me, had gone to join her kindred in the world above; for it is said that death is in search of a wreath, and he always chooses the fairest flowers. Perhaps friends and kindred, all had gone, and I had returned to view the forsaken hearth, and then go forth a lone and lonely wanderer, upon life's ocean, to float from shore to shore, without one guiding star, to lead me on my trackless way.

Such were the thoughts and feelings which were at work in my bosom, as I took the path to ascend the hill on which my former residence was situated.§ I chose a circuitous route to avoid passing through the town, for I was so covered with rags and tar that I felt ashamed to be seen by any civilized being. The clothing which I then had on, I had been obliged to wear for a month without the washing of a single article. I wore an old south-wester on my head, composed of canvass and tar; and on my body a woollen jacket darned and patched with an old blanket; and my pantaloons made of a blanket were patched in the knees and seat and sides with tarred canvass. My shoes were gone at the toes, and over the heels hung the remnants of a pair of stockings. I was truly a despicable looking fellow, and should have been a fine figure to sit for a painting of "the prodigal son."

Fearing lest I should meet some strangers, I went to the back door of the house and knocked.

The old woman who officiated as cook, and who had been long in the family, came to the door, and opening it started back in amazement. "Odds! life! Who's this?" she exclaimed.

"Why, Nabby, don't you know me?"

After looking at me for some moments, in utter astonishment, she exclaimed, "Why, Ben, is that you?" "Hush! hush! Nabby," said I; "don't speak so loud; for I don't wish any one to see me in this plight."

---

§The Varnum House sits high on the ridge overlooking East Greenwich harbor.

She gave me time to say no more, but started for the dining room, and called one of the young ladies out, who on seeing me threw back her arms and exclaimed at the top of her voice, "Why! Ben Ely!" The rest of the family, on hearing her exclamation, came rushing out, with several strangers; and there I stood surrounded, with all eyes turned on me, as if I had been some wild animal, for all were so lost in amazement, that they never thought of shaking hands; and had any thing affectionate stirred within them, the scent of tar and bilge-water must have sent the feeling back whence it came. After composing themselves a little, the first proposition was to feed the amphibious animal; the next to put him into a tub of hot water, and scrub him well; and the third to send for the tailor. For many days the scent of bilge-water could not be removed from my hair: and for weeks I could not get off my sea legs.

The pleasure which I experienced on returning home, afforded some compensation for many severe trials during my long and disagreeable voyage. Let me assure any young friend who wishes to become a sailor, that a life on the ocean wave is generally one of many hardships, and of few bright prospects. It demoralizes most persons who devote themselves to it; and raises but a few to nautical eminence, honour and wealth. Generally they who live on the sea have but a short life; and nine hundred and ninety-nine out of a thousand of them are but poor Jack Tars at last.

That navigation and commerce enrich and

aggrandize a nation none can deny; but what do most mariners profit themselves or their wives and children by all their voyages? The American sailor I think, may generally claim pre-eminence over mariners of other nations; but even the seaman of the United States is generally treated as if he were a slave, and abused in a worse manner than any favourite brute. All other classes of useful people seem to prosper more than sailors. Farmers, mechanics and tradesmen, by thousands have smiling families, domesticated in houses which they may call their own: but where will you find a sailor who has the fee simple of any house, except at the bottom of the ocean? Who, even among the captains of vessels become enriched by their frequent separations from their friends, and their long services on the deep?

To young men who long for hardships, and never expect to be any thing good, great, or useful, I would say "go to sea;" and perhaps I should say to a father, who can do nothing else with a worthless refractory son, "send him on a whaling voyage."

Thanks to a kind Providence, I am alive, and at home, and have had enough of a sailor's life. Whenever I meet a generous hearted tar, I shall have a fraternal affection for him; and pity him as one that will never have a name known among men; nor experience much of comfort; nor have a grave of his own; but I bid the ocean my long adieu. Poor, Old Joe! farewell.

Although I have spoken of Greenwich, R.I.,

as home, because it was my last place of residence before going to sea, and the abode of near and dear friends; yet in truth the home of my childhood was on the prairie at West Ely, Missouri; there is my mother's grave; and I beg leave to close my narrative by presenting some lines composed by myself at the mast-head of the bark Emigrant, while in the Mozambique channel, which I have entitled

THE LAND OF THE WEST.

'Tis the land of the West, 'tis the land of the West,
The scene of my childhood, the sweet home I love best;
With its tall forest trees, by the wild rose entwin'd,
Where the deer leaves the arrow in swiftness behind.

But it is not the shade of the tall forest trees,
Nor the voice of the birds, nor the hum of the bees,
Which gives to this land its most beauteous bright smile,
And calls loud to my memory o'er many a mile.

For many are the lands with their birds and their bees,
With their roses in clusters, and their hale green trees,
Whose mountains of granite arise towering in pride,
Their grandeur diffusing o'er vale and o'er tide.

But none of them boasting of fair prairies can tell,
Or the old Neptune's glories can rival so well,
For far as the eye o'er expansion can hover,
The mantle of sight seems an ocean to cover:||

||In his autobiography Ely makes a similar comparison between the sea and the prairies of Missouri. Speaking of the days of his early childhood, he writes: "West of the town of West Ely where our house was located, what was known as the Grand Prairie stretched away for thirty miles with no house or sign of inhabitants.

An ocean whose waves are of grass and of flowers,
Redolent with fragrance from golden hued bowers;
An ocean, whose islands are the wheat fields so bright,
Whose harbors, for light-houses, have cottages white.

The bright lights within are brilliant eyes of the fair,
And the spirit of love the combustible air:
Ignited they are gleaming far, far out to sea,
And invite the lone sailor boy home to their lee.

There each month hath its plant, each plant its own flower,
Called forth and perfected by each sunny hour;
And as the blue sea hath its gold dolphins and spray,
So the prairies resemble the rainbows of May.

Like the stars in the waters, there you see the bright bowls
Of the Indian plant, pointing its leaves to the poles,
While close by, the wild plume with its crimson tipp'd
head,
Stands erect in the midst of the violet's bed.

There the deer dips his hoofs in the strawberry's die,
And the quails in great coveys no longer are shy;
And the prairie hen feasts on the tall standing grain,
And the sheep are as white caps that fly o'er the main,

I love it, I love it: to my heart it is dear;
E'en in the Mozambique I still feel it near:
'Tis the land of the West, 'tis the land of the West.
The scene of my childhood, the sweet home I love best.

THE END.

To mark the roadway leading to the next settlement a single furrow was made by a prairie plough to the first settlement lest travelers who sought to cross it might be lost in the tall grass. In traveling the broad expanse it seemed as though one were at sea out of sight of land."

Appendices and Principal Sources

APPENDIX I

Conversion on a Whaleship

The following is Ely's own account, from his manuscript autobiography, of his religious conversion on board the *Emigrant*. Though he afterward for a time backslid and on his return from the voyage turned first to the law rather than the ministry, he never forgot the experience and refers to it repeatedly through the rest of his autobiography. Indeed, he tells us that it and his dying mother's prayers were what eventually, years afterward in California, caused him to seek ordination and to spend the remainder of his long life as a Presbyterian minister. This chapter is in the tradition of the numerous nineteenth-century religious tracts telling of the conversion of sailors; it can be compared to the conversion passages in James Fenimore Cooper's *The Sea Lions* (1849). It has intriguing parallels also to Samuel Taylor Coleridge's "The Rime of the Ancient Mariner," in which the rain of heavenly Grace descends only after the seaman has thrilled to the beauty of the water snakes; and to the masthead passages in *Moby-Dick*.

The Indian missionary John Bemo, whose influence on Ely is so plain in his story, was indeed an unusual man. According to one authority, he was "one of the most remarkable Indians of his time. His early life was filled with adventure, and in the absence of complete and accurate records, probably cannot be reconstructed. In its main outlines, however, his story appears in the records of the Seminole missions. He was a nephew of Osceola, and belonged to an Indian family who lived in the vicinity of St. Marks. A French sailor, Jean Bemeau, picked him up when he was ten or twelve years old and took him on a sea voyage. He served as cabin boy, became a first-class sailor, and learned something of ship carpentry. As a young sailor, he came into contact with Reverend Orson Douglas, pastor of the Mariner's Church in Philadelphia. Impressed by the frank manner and bright mind of young Bemo, the minister took steps to obtain schooling for him; and in the course of his studies, the Seminole lad became convinced that he ought to devote his life to the education of his people. In 1842, Orson Douglas planned for him to be sent to the new home of the Seminoles in the West, and on November 1 of the following year, he reported to Agent Thomas L. Judge. Bemo became acquainted with Harriet Lewis, a young Creek woman who had interests similar to his own, and they were married. Perhaps no Indian teacher had a greater influence for the adjustment of native people to white men's culture than Orson Douglas's protégé, John Bemo."*

Bemo's salary for teaching school at Prospect Hill in the Creek Nation, was three hundred dollars a year. In 1932 his descendants were still living near Muskogee, Oklahoma. Ely must have heard Bemo speak in 1842 or 1843, shortly before the latter's departure for his mission.

---

*Edwin C. McReynolds, *The Seminoles*, page 256. Copyright © 1957 by the University of Oklahoma Press. Quoted by permission.

that one should rise from the dead. I thought to myself, "That's it! — that's it — that is the end of controversy." With no more reliable foundation for my skepticism I tried to believe that the Bible was not reliable and that the God of the Bible did not exist. The real reason for my [opposition?] was that the Bible and the thought of a personal God to whom I was accountable were antagonistical to my feelings and desires. My experience was a practical demonstration of the truth that the carnal mind is [in] enmity against God — that it is not subject to the law of God, neither indeed can be.

During the time I was a pupil at Lawrenceville our school was visited by John Bemo, a Seminole Indian, the nephew of the celebrated chief Osceola, who was going to the Indian Nation as a missionary. In an address to the School he gave an account of his conversion while at sea. He said in substance that when a small boy his father took him to Saint Augustine, Florida, and that whilst there his father was killed in a drunken fight with another chief. He said that whilst wandering around destitute, homeless and friendless, some sailors took him aboard their vessel and carried him to sea. Some time afterward, presumably after he had acquired some knowledge of the Christian religion, he became deeply impressed with a sense of his sinful and lost condition. One day whilst thus deeply depressed, looking over the side of the ship, he saw a dolphin change his colors. He said that the sight was so wonderful and beautiful that it suggested the question — Who made the dolphin? and the answer came — the Great Spirit, followed by the thought — If the Great Spirit could make the dolphins, He can take the burden off my heart. He said that he went down into the forepeak of the vessel and prayed and that, while praying, peace came to his soul. When I heard these statements, with a self-satisfied assumption of my own superior wisdom and condescension, I thought to myself that though the Indian was sincere his statements were the result of ignorance and superstition. I had no idea that, months afterward, his story would be the means of leading me to seek peace by imitating his example.

It came about in this way. I had tried to believe that there was no God; but there were thin places in the clouds, where the truth that there was a God and that the Bible was His inspired word would break through. This stirred up within me a spirit of unrest and rebellion. I became miserable. At times in my desperation even in the midst of a storm as the thunder rolled and the lightnings flashed I have shaken my fist at the heavens and said, "If there is a God, I hate him! If there is a God, I wish he would strike me dead!"

After months of anxious thought, I was finally compelled, I think I may say, against my will to believe that there was a God and that the Bible was His Word. I saw the unmentionable obscenity, vulgarity and degradation that rendered a whole ship a floating hell. Even the names of mother and sister were not too sacred to protect them from obscene language, and though I was terribly profane myself I heard blasphemy in the forecastle and from the lips of the officers that made me shudder, and when I visited the Island of Madagascar and saw the degradation of the heathen — saw brothers bring their sisters to the ship and offer to sell them for a bolt of calico, or a piece of iron hoop, for the purposes of prostitution — the question was forced upon me — Why are these sailors [*crossed out:* shipmates of mine] and heathen so degraded and the people of Christian lands and the society with which, at home, I associated so far superior?

I could find but one satisfactory answer, viz., that it was the influence of the Bible and Christian civilization. I thought of the writings of the heathen philosophers, of Plato, Socrates, Confucius and Seneca, which though they contained many good things, had failed to secure the moral results procured by the Bible, and I was convinced that the Bible therefore must be the word of God. I did not at first yield to my convictions, but one morning as I was sitting at the masthead thinking of these things I smelt the sweet breath of the land breeze which at sea bears the odor of a newly mown hayfield. It reminded me of my boyhood's home — of my father — of the dying prayer of my beloved mother, and as these sacred memories were called to my recollection I felt that I was a poor prodigal who had wandered away from my earthly and heavenly father — that I was an outcast. My heart was saddened and softened. Just then a dolphin came up alongside of the ship and as he did so he changed his colors, irradiating the different rays of the rainbow, and as I looked upon the beautiful transformation it reminded me of the story of the Indian missionary which I heard when a schoolboy at Lawrenceville, New Jersey, and I said, "I will try the Indian's plan — I will pray to God." Just then I was relieved from my watch aloft and went below. After turning into my bunk, I began to pray, and taking from my chest, which was lashed beside my bunk, a Bible that had been given to me by a young lady friend just before I went to sea, I happened to open it at the 15th Chapter of Luke. As I read the story of the Prodigal Son, I said to myself, "That is just my case. I am a poor miserable prodigal far away from my Father's home."

BILL OF HEALTH.

United States of America.

DISTRICT OF BRISTOL AND WARREN.

TO ALL TO WHOM THESE PRESENTS SHALL COME: We the Collector and Surveyor of the Port of *BRISTOL AND WARREN*, do by the tenor of these presents, certify and make known, that the Captain, Officers, Seamen and Passengers of the Barque called the Emigrant of Bristol laden with Provisions & Apparatus for carrying on the Whale-Fishery

and of which James Shearman is Captain; consisting of twenty-one Officers and Seamen, and no Passengers, now ready to proceed on a voyage to the Indian Ocean and elsewhere beyond sea, are all in good health.

AND WE DO FURTHER CERTIFY, THAT there is not at this time any contagious disorder nor Cholera prevailing in this port or its vicinity.

GIVEN under our hands and the seal of the Custom House, at Bristol, the ninth *day of* November *in the year of our Lord one thousand eight hundred and forty-*four *and in the Sixty-*ninth *year of the Independence of the said States.*

The Bill of Health issued to the *Emigrant* on November 9, 1844; one of her required clearance papers. (Courtesy Rhode Island Historical Society)

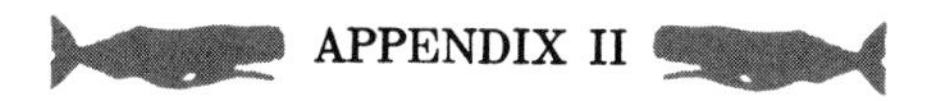

APPENDIX II

The Vessel, the Voyage, and the Crew

I. The Vessel: The *Emigrant*

WHEN young Ben Ely sailed his yacht from East Greenwich across Narragansett Bay to Bristol to sign on a whaler, the vessel fitting out in that trim little whaling and slaving port was the *Emigrant.* Built in Swansea, Massachusetts, in 1833, she was then eleven years old. According to the registers, she was 85′ 6″ in length, 22′ 10″ in beam, and 10′ 4″ in depth. Her burden was listed as 178 38/95 tons. She had been originally rigged as a brig, with two masts, but in 1841 she was altered to bark rig, with three masts. She had a billethead and square stern. Her first registry, October 23, 1833, had listed her hailing port as Warren, Rhode Island, as had also her second, on May 18, 1840. But in 1841 she had been bought by members of the prominent Church family, shipowners of Bristol, and the owners when Ely signed on were Samuel W. Church, Thomas Church, Stephen

Church, William H. Church, Josiah R. Talbot, and Ephraim Gifford, all of Bristol. Of these Samuel W. Church was the principal. Under these owners she had cruised for oil in the Atlantic from August, 1841, to June, 1842, Samuel Lake master; from June, 1842, to January, 1843, Peleg Tripp master, coming back "clean" (without oil); and then under James Shearman, Ely's captain, from February 8, 1843, to September 9, 1844.

After the bark's return from the cruise narrated in *"There She Blows,"* she sailed with a crew of only eight (she shipped twenty-three for a whaling cruise) on three voyages as a merchantman. One was to Cuba, one to St. Kitts, and the third ended almost as soon as it began when she was driven in distress into Boston and her crew discharged there. On June 1, 1848, under new owners and with her hailing port changed to New Bedford, she set out on what was to be her last and catastrophic whaling voyage. According to the *Whaleman's Shipping List* of October 16, 1848, she was found floating bottom up, stern and rudder gone, off French or L'Espérance Rock in the Pacific Ocean northeast of New Zealand. "Nothing is known of the crew," the account states, "and it is feared they perished." In poor condition and leaking badly even on Ely's voyage, she had evidently broken up and capsized in a fierce gale encountered by several vessels in the vicinity at that time. Though at first it was hoped her boats had managed to make safe landfall on some nearby island, no news ever came, and presumably the whole crew was lost. It

GENERAL CLEARANCE.

DISTRICT OF BRISTOL AND WARREN, ss.

PORT OF BRISTOL AND WARREN,

THESE ARE TO CERTIFY all whom it doth concern, that James Shearman master or commander of the Barque Emigrant burthen one hundred seventy-eight 38/95 ths tons or thereabouts, mounted with no guns, navigated with twenty-one men, United States built, and bound for the Indian Ocean & elsewhere having on board Provisions and Apparatus for carrying on the Whale Fishery

hath here entered and cleared his said vessel according to law.

GIVEN under my hand and seal, at the CUSTOM HOUSE, *at Bristol, this* ninth *day of* November *One Thousand Eight Hundred and Forty* four. *and in the Sixty* ninth *year of the Independence of the United States of America.*

The *Emigrant's* General Clearance, November 9, 1844. Originally among the papers of the Bristol-Warren Custom House, now held by and reproduced courtesy of the Rhode Island Historical Society.

was well for Ely that he had not waited to ship for this cruise.

II. The Voyage: November, 1844 – February, 1847

But what about the voyage recounted in *"There She Blows?"* Where did it go? How long was it? What success did it have? Though the most interesting account is that of the book itself, from the Bristol Customs House records preserved in the library of the Rhode Island Historical Society in Providence, from entries of ships sighted in the *Whaleman's Shipping List,* and more especially from surviving manuscript journals by two other crewmen on the same voyage, fascinating factual data and corroboration can be adduced to make more vivid Ely's imaginative and impressionistic account. By surprisingly fortunate chance both the journal kept by the ship's cooper, Charles F. Tucker of Fall River, and that kept by the one American boatsteerer, Seth F. Lincoln of Bristol, have been preserved. Both give day-by-day information about the voyage, as may be seen in Appendixes III and IV of this volume.

On November 9, 1844, in preparation for its departure, the *Emigrant* was given its Bill of Health and its General Clearance, the latter certifying that "James Shearman master or commander of the Barque Emigrant burthen one hundred seventy-eight 38/95ths tons or thereabouts, mounted with no guns, navigated with twenty-one men, United States built, and bound for the Indian Ocean and elsewhere having on board provisions and apparatus for carrying on the Whale Fishery hath here entered and cleared his said vessel according to law." The Sea Letter, a long and impressive document in four languages (English, French, Spanish, and Dutch) addressed to the kings, emperors, archdukes, barons, and other notables of the world, signed in facsimile by J. Tyler, President of the United States, and Abel P. Upshur, his Secretary of State, gave the bark permission to depart and requested protection for it on its voyage. A Report and Manifest of Cargo had also been made out, averring that the vessel carried only "Sea stores and apparatus for carrying on the Whale Fishery." She carried provisions for two years; Seth Lincoln listed them in his journal, and that list is reproduced here.

LIST OF PROVISIONS ON BOARD
NOVEMBER 12th 1844
(From Lincoln journal)

[NOTE: This list is especially interesting in the light of Ely's charge that Captain Shearman starved his crew.]

Meat 80 blls Beef 40 blls Pork	120 blls
Hams 200 lbs	200 lbs
Flour 40 blls	40 blls
Bread 10:000 bbs	10:000 bbs
Butter 5 kegs 100 lbs each	500 lbs
Cheese 10: 25 lbs each	250 lbs
Apples 3 blls 100 lbs each	300 lbs
Pickles 1 bll	1 bll
Raisins 1 keg 25 lbs	25 lbs
Sugar 2 blls 250 lbs each	500 lbs
Molasses 11 casks 90 galls each	990 galls
Coffee 2 bags 160 lbs each	320 lbs
Tea 2 chests 50 lbs each	100 lbs
Rice 2 blls 200 lbs each	400 lbs

Bought at Mauritius May 20th 1845

Beef 6 blls	6 blls
Flour 2 blls	2 blls
Bread 400 lbs	400 lbs

Bought at Nos Beh July 12th 1846

Bread 1000 lbs	1000 lbs
Rice 6 bags 50 lbs each	300 lbs

Provisions remaining on board

Meat 4 blls Beef 3 blls Pork	7 blls
Flour 1½ blls	7 blls
Bread 500 lbs	500 lbs
Molasses 45 galls	45 galls
Coffee 20 lbs	20 lbs
Rice 40 lbs	40 lbs

With these preliminaries over, at eight o'clock in the morning of November 10 the *Emigrant* weighed anchor at Bristol. She proceeded out Narragansett Bay toward Newport but — ill omen for the start of a cruise! — struck on a rock in Newport harbor and wasted two hours getting herself off, not reaching port till 6:00 P.M. On the eleventh she cleared Newport and was off on her long voyage.

Her first port was Brava in the Cape Verde Islands, which she sighted on December 17 and where on the eighteenth Captain Shearman went ashore for recruits. Then she was off again to the right whale grounds of the South Atlantic. On January 3 she crossed the equator and, as Ely tells, Neptune came aboard to shave the uninitiated. By January 26 she was far to the southwest, in the La Plata estuary, Argentina. Then, cruising for right whales as she went, she set sail almost due east. On March 30 she raised the isolated island peak of Tristan da Cunha, and on the following day sent a boat ashore for potatoes and other vegetables. On April 4 she crossed the meridian. By the ninth she was some 218 miles off the Cape of Good Hope. Passing the dangerous Agulhas Banks, she sailed on into the Indian Ocean, sighting Réunion (Isle of Bourbon) on May 13. Because of an error in their chronometer, the *Emigrant's* crew at first took it to be Mauritius (Isle of France). On the fifteenth they left the coast of Réunion and on the next day reached the real Mauritius. On the seventeenth they were towed by a steamboat into the harbor of Port Louis, where they repaired the damage suffered during the hard gales of April and where the men were given liberty. Four seamen took the opportunity to desert ship and were here replaced by four others. Not until June 6 did the *Emigrant* leave port and head northwest for Madagascar. On June 21 they saw the great island for the first time. On June 24 they doubled Cape Ambre, its northernmost tip, and sailed into the Mozambique Channel. On the twenty-sixth they were off Cape St. Sebastian, and on the twenty-seventh they dropped anchor at the important whaling station on Nos Beh,* an island just off the western coast. This was to be the *Emigrant's* main base.

*In the names and spellings of the places on and around Madagascar I have followed the 1841 edition of Horsburg's *India Directory* rather than modern gazeteers. These were the names the whalers themselves knew. The voyage can be followed in more detail by reference to the excerpts from Tucker's and Lincoln's journals.

Now on her chosen grounds, the bark cruised for more than a year around Madagascar, spending most of her time in the Mozambique Channel between the island and Africa. Many were the exotic places she visited — though well known to the whalemen of those days. On July 10 they were off Nouk How Island, on July 12 near the Chesterfield Banks, and on August 2 at Juan de Nova Island. They then returned to Nos Beh for two weeks (October 8–24), going across Passandava Bay for wood and water and to French Town (probably what is now Hellville) for shore leave. On November 2 they set sail again and by November 20 were off Cape St. Mary on the southern tip of Madagascar. From March 7 to 22, 1846, they anchored in St. Augustine's Bay to refit, obtain wood and water, and give the crew liberty. In May and June they were cruising near Sandy Island (now Tromelin) 260 miles off Madagascar's northeast coast. On June 18 they doubled Cape Ambre again and ran down once more to Nos Beh, where they left two men whose time was up. Later they sailed up Majambo Bay on the west coast (June 28–29) and anchored off Manjunga Point (June 30–July 1). By July 3 they were back again in Passandava Bay near Nos Beh, where they got wood, scraped and painted ship, and, in French Town nearby, picked up more recruits. On July 9, 12, and 15, the men had liberty in French Town. Then on July 22, after the ship had been cold-tarred, they set sail for a cruise along the African coast. From August 17 to 28 they were off Juan de Nova, then from August

30 to September 4 again near the African coast, this time close enough to sight (September 3) Fogo Island and Fire Island (probably two names for the same island). Once, at least, they circumnavigated the whole island of Madagascar, rounding Cape St. Mary, the southern end, and cruising off Fort Dauphin on the eastern coast before sailing northward to Cape Ambre. Their cruises made an intricate pattern all around the island — but after a little early luck they got no oil.

Finally even the captain's grim patience evidently wore out. After cruising vainly in the Mozambique Channel, sometime toward the middle of September, 1846, hoping to capture right whales if he could not kill sperm, he set the course out of the Indian Ocean into the South Atlantic. He cruised there for several weeks. But, despairing of more oil, with his supplies low, and perhaps concerned over the leaking of the bark, Shearman at last set the course for home. On December 6 they raised St. Helena and on the seventh passed to the eastward of it. They sighted Ascension Island on December 15 but did not land because no water was available there. December 22 to 27 they spent near the Equator, but with the new year (1847) turned directly homeward. By January 21 they were off Cape Hatteras; on January 25 the change in the weather and water told them they had crossed the Gulf Stream; and on the twenty-seventh they were southeast of Montauk Point when a gale caught them. Two days later they saw Block Island and took aboard a pilot, and at

6:30 P.M. of the same day (January 29) they dropped anchor off Fort Adams in Newport harbor. But even then their troubles were not done. Because of a fierce gale they had to haul in their anchor and were not safely berthed at Newport until February 2. On February 3 they sailed up the Bay and came at last to rest at Bristol. They had been gone two years, two months, and twenty-five days.

During the voyage the *Emigrant* spoke a number of vessels, most of them whalers. According to a careful list in Tucker's journal, she had spoken the brig *Edward* of Mattapoisett (January 27, 1845); the bark *Superior* of Sag Harbor (January 29); the bark *Bruce* of Fair Haven (July 9); the bark *Hope* of New Bedford (November 25 and December 27, 1845; January 25 and March 31, 1846); the bark *Lafayette* of Warren (January 1, 1846); the bark *Charleston Packet* of New Bedford (February 12); the ship *Susan* of New Bedford (March 8); the bark *Frances* of New Bedford (April 2); the bark *Hector* of Warren (July 21); the brig *Ann* of Liverpool — a merchantman (October 14); the ship *Waverly* of New Bedford (November 10); and the bark *Express* of Wiscasset — a merchantman (January 23, 1847). Whether on the River La Plata or at Nos Beh or off Fort Dauphin or in St. Augustine's Bay or at other far ends of the earth, the New England whaleman was in one sense never far from home and neighbors. Indeed, rarely throughout the whole voyage did more than a few days go by without one or more sails being seen.

But the *Emigrant* was not cruising for adventure

or to meet friends; she was hunting whales. What success did she have? The cruise started well. Blackfish and porpoises where taken in the Atlantic. On February 4, 1845, and again on the seventeenth, the boats successfully lowered for right whales. Then from June to December, 1845, nine sperm whales were captured: one on June 13, one on August 20, one on September 6, and two each on September 10, September 11, and December 5. But with the end of 1845 luck failed. Throughout all of 1846 there were only twelve sightings, and not one of these resulted in a capture, though several times the boats got fast. The *Emigrant,* provisioned for only two years, grimly stuck to the chase, much to the discomfort and expense (as Ely tells) of the men, who had worn out their supply of clothes and were forced to buy at high prices from the slop chest. In his log Tucker reflects grimly the rising tension.

April 25, 1846: "4 mounths this day sence we have seen the Spout of a Spirm Whale."

May 3, after they have killed a sperm whale only to watch him sink: "So much for working on the Sabbath."

August 5: "8 months today sence we have taken a Whale."

August 10: "21 MONTHS OUT TODAY WITH 300 SPERM AND 135 OF WHALE OIL."

September 5: "9 MONTHS TODAY SENCE WE HAVE TAKEING ENEY OIL."

September 10: "This day 22 months out with 300 bls of Sperm Oil and 135 bbls of Whale Oil."

October 5: "Have taken no oil for the last ten months."

October 10: "23 MONTHS OUT WITH 300 bls of sperm and 135 of whale OIL,"
November 5: "Have not takeing a drop of oil for the last eleven months."
November 10: "Two years out today with 300 bbls of Sperm oil and 135 bbls of Whale Oil."
December 5: "12 Months today scence we have takeing eney Oil."
January 5, 1847: "13 Months today scence we have takeing eney Oil."
January 10: "26 Months out today with 300 bls of Spirm and 135 bls of Whale Oil."

Lincoln's comments, in contrast, have a certain ironic humor. On November 12, 1845, speaking of the first year, he says; "This day completes the first year of our cruise. We have taken seven spirm whales which made us 220 blls of oil and 2 right whales which made us 140 blls making in all 360 blls." Then he adds, whether ironically or not I do not know, "Great doings indeed. Fortune has indeed favored us so far." But if it is uncertain in November, the irony is clear enough by April. On the twelfth of that month he writes: "And so ends this day. 17 months out this day. We are on the road to fortune." And even more wryly on December 10: "Twenty-five months out this day. Fortune has indeed favored us so far and if she continues her favors a few more years we shall probably end our days in the almshouse or states prison. But let us hope on." It was no fun to be on a whaling ship that could not catch whales.

When at last on February 2, 1847, the *Emigrant* reached Bristol, the Return of Merchandise unladen under the inspection of Allen K. Munro, Inspector, and Nathan Bardin, Surveyor of the Port of Bristol, was as follows:

> 300 barrels or 9450 gallons of sperm oil valued at $9450
> 150 barrels or 4725 gallons of whale oil valued at $1775
> 1400 pounds of whale bone valued at $500

The total value of the voyage was $11,725. Besides, there were "empty casks" (of which there must have been many) and "whaling apparatus." The Churches, the *Emigrant's* owners, probably made expenses and a small profit (one would like to see the account book), though it was not a successful voyage. For the crew, as Ely indicates, it was a disaster. "Old Joe," one of the seamen, after more than two years of privation and danger, was paid off with $13; Ely himself makes no mention of being paid at all. But for the common sailor almost every whaling voyage was a disaster. Charles F. Tucker's accounts, showing his receipts and expenditures on the voyage, are illuminating in this connection. They are reproduced on the following page.

CHARLES F. TUCKER'S ACCOUNTS
ON BOARD SHIP
1845–1846

Money received from the Ship

Isle of France	— May	$4.00
Nos Beh	— July	1.00
do	— Oct.	1.00
do	— July	.50

[Expenditures on board]

July 31	jack knife	$ .50
Feb. 1	15½ pd of Tobacco — 25 cts pd	3.87
Oct. 23	20 yds cotton cloth	2.80
do	straw hat	.30
Dec. 23	two pr duck pants — 1.10 pr	2.20
do	Sheath nif 35 thread 15	.50
do 26	pair Slippers 40 do 33	.73
May 23	jack knife	.33
July 21	1 doz of pipes	.12½
do	Dates	.50
Oct. 2	Stockings 42 cts thread 6	.48

[NOTE: The cotton cloth and much of the tobacco were probably used for trade with the natives. See Ely's comment in the book and in his account of his conversion about brothers who "bring their sisters to the ship and offer to sell them for a bolt of calico, or a piece of iron hoop, for the purposes of prostitution."]

III. The Officers and Crew

Especially for the man before the mast, the character of a cruise may be more dependent on his officers and shipmates than on the number of whales killed or even the ship. Ely himself gives the most vivid description of the men on the *Emigrant*, but his impressions can interestingly be supplemented from facts drawn from the actual crew list and other records concerning the officers and crew.

The captain of the *Emigrant* on Ely's cruise, as he had been on the vessel's preceding cruise, was James Shearman, Jr., of Fall River. Born at Somerset in 1807 or 1808, he had shipped at eighteen, probably as second or third mate, on board the whaling brig *Charleston Packet*, by chance one of the vessels that the *Emigrant* spoke in the Indian Ocean. For a number of years he was mate and then captain of the ship *Young Phenix*, in which he owned a share. After two voyages (1843–1844, 1844–1847) in command of the *Emigrant*, he was in 1848 listed master of the *Timoleon*. Then he disappears from the whaling scene. He seems to have lived for a while in Sidney, Maine, on the side of the Kennebec. A copy of a will in the Bristol County, Massachusetts, probate registry seems to indicate that he died at Sailors' Snug Harbor, Staten Island, though no record of him remains there. In 1844 he was thirty-six years old, 5 feet 11½ inches tall, with brown hair and light complexion.

The mate was Samuel B. Allen, born in Tiverton, Rhode Island, in 1820. He was 5 feet 8¾ inches tall, with light complexion, brown hair, and blue eyes. He had received his protection certificate in 1838. He had been mate with Shearman on the *Emigrant's* 1843 voyage. Later (1850–1852) he was mate of the ship *Brighton*.

LIST OF PERSONS COMPOSING THE CREW
of the Barque Emigrant of Bristol whereof is Master James Sherman bound for Indian Ocean & Elsewhere Whaling

Names	Places of Birth	Places of Residence	Of What Country Citizens or Subjects
James Sherman	Somerset Mass	Fall River Mass	United States
S B Allen	Tiverton RI	Fall River	do
Thomas Palmer	Westport Mass	do	do
Seth F Lincoln	Bristol RI	Bristol RI	do
Emanuel Joseph			
Joseph Hammond			
Albert R Barnes	Bristol RI	Bristol RI	United States
Benj[n] E S Ely	Philadelphia Pa	Philadelphia Pa	do
John Joseph			
Joseph Antone			
Andrew Borden	Fall River Mass	Fall River Mass	United States
Charles F Tucker Jr	Portsmouth NH	do	do
Seth Cole	Warren RI	do	do
Oliver Arnold	East Greenwich RI	East Greenwich RI	United States
Matthew Gill			
Christopher McHale	East Greenwich RI	East Greenwich RI	United States
Antone Eigo	Pensacola Florida	Bristol RI	do
James A Hall	Newport RI	Newport RI	United States
Joseph Knights	Boston Mass	Boston Mass	do
William H Saunders	Glocester RI	Glocester RI	do
William P Chesley	Londonderry NH	Providence RI	do
	[ADDED LATER]		
William K Dupee,	Boston,	Boston,	United States,
Bernard McClusky,	New York,	so delivered but no proof offered,	
Seamen Johnson,	New York,	New Bedford,	
Henry Henry	Cambridge,	Boston	United States

Description of Their Persons

Aged	Height feet	Height inches	Complexion	Hair
26	5	3¾	Dark	Dark
22	5	8	Light	Brown
17	5	5	Dark	Black
17	5	5½	Dark	Black
15	4	8½	Light	Light
27	5	4½	do	Brown
17	5	4¼	do	Dark
18	5	10	Light	Brown
19	5	1	Light	Dark
22	5	8	Dark	Black
25	5	4½	Light	Light
34	5	6¼	Dark	Dark
19	5	10¾	do	Brown
21	5	8	Light	do
24	5.	6½	Brown,	Brown,
22	5	11½	Dark,	Brown,
24	5	6¼	Dark,	Brown,
29	5.	7	Black,	Woolly,

Ely, who admired Allen, was even more enthusiastic about the second mate, Thomas D. Palmer of Westport. He calls him "in grain one of nature's gentlemen." Palmer was twenty-six, 5 feet 3¾ inches tall, dark hair with dark or ruddy complexion. He had served on the bark *President* in 1837–1838. A few days after his return to Bristol in the *Emigrant,* he shipped out again as mate of the *Mexico.*

The Crew List of the *Emigrant* (which I have reproduced in these pages) adds detail to Ely's descriptions of his other shipmates. Here appear the names of Seth F. Lincoln, boatsteerer, and Charles F. Tucker, cooper, the writers of the two extant journals and both mentioned in *"There She Blows."* According to Ely, the other two boatsteerers were "Manuel, a native of Portugal" and "a black Joanna man, of the race of pirates." The former is identifiable on the list as Emanuel Joseph. The other, evidently of Malay stock and from Johanna, one of the Comoro Islands near Madagascar, probably signed on as Joseph Hammond, English names often being assigned foreign sailors. On the list too is Seth Cole, who succeeded Ely as steward and who was disliked by the crew as a pet of the captain. Not suprisingly, no name on the list corresponds with that of Ely's cook, James Sables, who according to Ely's account was accompanied everywhere on board by his pet pig. Alas, he is probably a mythical figure. But the black assistant cook can probably be identified with Henry Henry of Boston, the only black-complexioned man with woolly hair on the list. Surely Mickey the Irishman is Matthew Gill, and Old Joe was Joseph Knights of Boston, aged thirty-four. That was "old" for a seaman on a whaling vessel. In addition to mate Allen, five of the crewmen had sailed on the *Emigrant's* previous voyage with Shearman.

Though there were surprisingly few desertions, the *Emigrant* like most whalers did not return to port after her long cruise with the same company with which she set forth. In his book Ely says that six men deserted and that two beachcombers were signed on to fill their places. Tucker says that four men jumped ship at Port Louis and that four replacements were signed on there to fill their berths. He also mentions two men picked up just for the cruise at Nos Beh and returned there when their time was up, only to be replaced at nearby French Town with two others. This casual coming and going of temporary recruits is not recorded on the crew list, but even on it some change of crew is shown. According to a statement sworn to June 4, 1845, by Captain Shearman before H. Hollier Griffiths, United States consul on the island of Mauritius, and vouched for again by the inspector of the port of Bristol on the vessel's return home, four men (William P. Chesley, Emanuel Joseph, James A. Hall, and Antone Eigo) deserted at Nos Beh about July 20, 1846. At such a busy whaling station they could, of course, easily find new berths. To take their places four men were signed on at Port Louis: William K. Dupee, Bernard McClusky, Seamen Johnson, and Henry Henry. McClusky, perhaps one of the term men, later left the bark elsewhere. Comparatively, this was a good record for desertions; by far the greater part of the *Emigrant's* company returned in the vessel. Perhaps Captain Shearman was not quite the tyrant Ely shows him. Many whalers lost more than half of their seamen in the course of a cruise.

The composition of the crew is instructive, too. Aside from the officers, who were themselves young, of the twenty-five men down on the crew list as having been aboard the bark on this cruise, only three whose ages are ascertainable were over twenty-five, and they were twenty-seven, twenty-nine, and thirty-four. About two-thirds of the crew were United States citizens. Most came from New England. Probably this preponderance of natives reflects a small whaling port such as Bristol, Warren, or Duxbury. Vessels sailing from New Bedford or Nantucket even at this era generally had a larger proportion of foreign-born.

To the reader of *"There She Blows,"* the most interesting entry in the crew list is naturally that for Ely himself. Here he is, summed up briefly: "Benjn E S Ely"; place of birth: "Philadelphia Pa"; place of residence: "Philadelphia Pa"; citizenship: "United States"; aged 17; height 5 feet 5½ inches; dark complexion; black hair. Since Ely's first name was Ben, the "Benjamin" was an error. Actually, too, he was only sixteen years nine months old, but seventeen perhaps sounded better.

APPENDIX III

Excerpts from the Journal of Charles F. Tucker, Jr., Cooper of the Emigrant

November the 10 [1844] remarks on bord: This day have ben employed in weaighing anchor and getting under way. got under way at 8 A M with a very light breeze from the N & W and part of the time ded calm arrived at New Port about 2 P M run on a rock and layed about 2 hours. got of and came to anchor at 6 o clock. so ended our first day

remarks on board tuesday the 12 this day ben employed in weaighing anchor and got ship under way about 8 A M and put for sea with a fine breeze from N & E and cloudy with a plenty of fog

remarks on bord the 19 have had very strong gale the past 2 days from the S W in wich we got our wast boat stove and foresail split and done some other trifeling damage

Dec the 18 at Bravo. one boat ashore after recruits and so forth

Dec the 31 . . . good wether and pleasent with light breezes lowered and got a black fish this afternoon. this evening saw a comet the tail barring to the northard

Sunday January the 26th [1845] on the river Le Plate good wether and pleasent with moderate gales

Feb the 2nd . . . at 4 PM lowered and struck a sun fish and took him on bord. nothing in sight but finbacks and a plenty of them

Feb the 4 . . . at 1 P M saw right whales and at ½ past 1 lowered away the boats and struck and got one. took him along side and comenced at 5 o clock a cutting him in. got him cut in at half past 1. at 2 got supper very early indeed

Feb. the 9 . . . the past week all hands have been boiling out. had a gale of wind the past week and had to cool down for one night. commenced trying the next morning got all tryed out friday and stowed him down yesterday saw a sail yesterday standing to the northard supposed her to be a merchantman

Feb the 13 [on the right whale ground] . . . saw whales yesterday and lowered but got none. nothing in sight today but finbacks

Feb the 16 . . . pleasent with strong gales from the southard have had bad wether the past few days fogey and rainey and strong gales

Feb the 17 . . . saw right whales this forenoon and lowered the boats and struck one got him alongside and commenced cutting at ½ past 5. cut until 9 o clock and then left him until morning. got supper and set the wach and then turned in so much for this day of our Lord

Sunday Feb the 23 . . . the past week got a whale tryed him out and stowed him down

Tuesday the 25 . . . this day saw and lowered for whales but got none

Thursday Feb 27 . . . saw and lowered for a right whale this afternoon and galled him and did not get him

March the 5 . . . saw a spirm whale the latter part of the day about 1 mile to windard did not lower the boats it being very ruged and he goin eyes out

March 30 . . . saw the land this afternoon at sun down about 80 miles from the land and layed aback until morning

March 31 at Tristedo Anccunah . . . stood in for the land and sent a boat ashore this forenoon the boat returened at night with potatoes and so forth at 7 squared the yards and was O P H

April the 4 on the Merredeon . . . strong gales from the Northard stearing E S E with a fair wind and flowing sheat goin at the rate of 10 not

April the 9 . . . of the pich of the Cape good Hope

Monday April the 14 . . . begins with a very strong gale from the W ard and a very hevy sea on and we goin ded before it under a close reefed maintopsail and foresail. such a mountainous sea on it is imposable to lay to — at 8 o clock A M a sea struck our larbord boat and carried it away and with it the daveys and boats falls and all the boat craft excepting oars at 9 another sea stove and carred away the starboard boat and daveys and the boats craft and stove in some of our bullwarks done up some other trifeling jobs

Sunday May the 4 . . . had strong gales the past week and had to lay to under a close reef maintopsail and foresail. turned up the starboard boat and took the craft out of the other boats redy to turn them up

Tuesday May the 6 . . . this day begins with strong gales from the E S E with a bad sea on and squaly at 10 A M a squal struck us and carred away our foretopmast and foretopgallant mast and royal mast and maintopgallant mast and royal mast and so forth and done up some of her other od jobs without leave or licence. all hands employed on clearing the wreck. got all cleared away and everything secure. at 7 P M took the waist boat in on deck and got every thing secure redey for the gale. by the looks of the wether we shall have one

Sunday May the 11 . . . 190 miles from the Isl of Bourbon

Tuesday May 13 . . . saw the land this morning and ran for it and followed it down till night and then stood of from the land

Thursday the 15 of May . . . left the Isl of Borbon yesterday for the Isle of France — made no stop at Borbon on the account of the chronometer being out of the way we made the wrong Island

Satterday May the 17 . . . ded calm at noon about 15 miles from the Citty at 1 P M the steam boat came to our releaf as it was ded calm and towed us in to the harbor where we now lay snug and secure

Sunday May the 25 . . . the past week have ben employed in painting ship and fitting riggen and so forth — one of the waches ashore every day the past week

Sunday June the 1 . . . all hands have ben employed in sending up the spars and riggen ship — and got one raft of watter — we are now reddy for sea — four of the foremast hands have deserted the past week and we have ship four more in their places

Friday June the 6 . . . Pilot came on bord at 5 P M we cast off the warps and went to sea

Thursday June the 12 . . . saw and lowered for Spirm Whales this forenoon but did not get fast — lowered again this afternoon and got fast — but did not get him it being ruged and a bad sea on and the Whale got so far to the windard that had to cut from him

June the 13 Friday . . . saw and lowered for spirm

Whales this forenoon and the Starbord boat got one — got him alongside and commenced at 3 o clock a cutting him in got him all in at 12 o clock and then got supper and set the waches — so ended this long — long — day.

Sunday June the 15 . . . commenced boilling yesterday and got through today

Monday June the 16 . . . have ben this day employed in stowing down oil

Friday June the 20 . . . bound for Nos Beh

Remarks on bord June the 21 . . . the land under our lee distent 112 miles

Sunday June 22 [in Mozambique Channel] . . . took in the foretopsail this morning. at 10 it moderate a little and put on the foresail and wore ship at 5 still continues to be squaley and blowy the old ship a jumping and pitching in it like a sick porpoise. a continued gale the last 7 days

Tuesday June the 24 . . . made the land at 1 o clock and stood for it until 5 clock when we closed reefed the topsails and stood of land distent about 12 miles

Wednesday June the 25 . . . run down with the land until dark and then tack ship and stood of land distent 5 miles

Friday, June the 27 . . . run with the land until half past 8 and then the wind died away and we came to anchor about 12 miles from the town [Nos Beh]

Satterday June the 28 . . . at 10 clock the wind haled and all hands caled to weigh anchor . . . . arrived in port and drop anchor about 4 clock

July the 4 — the day of indipendence . . . all hands employed about ship until 9 clock — we then had the rest of the day to ourself

Wensday July the 9 . . . all hands called this morning about 2 hours before day to weigh anchor and get the ship under way . . . spoke the Bark Bruce of Fair Haven 7½ months out — one hundred bls

July the 13 [bound on a cruise] . . . latter part of the day ded calm — with nothing but a sail in sight a French Man of War

July the 17 . . . the land in sight under our lee distense about 12 miles — the last three days have had strong gales from the Southard

Thursday July the 27 [off the coast of Africa] . . . finbacks humpback and jumpers in sight

Wednesday July the 30 . . . lowered the boats this forenoon for grampuses but did not get fast and again in the afternoon but did not get fast — ded calm

Friday August the 1 . . . lowered this afternoon for Blackfish but got none

Satterday Aug the 2 . . . saw land [Juan de Nova] this forenoon and run for it past to the leeward of it distent 4 miles very low land

Monday Aug the 4 . . . this day begins with fine breezes and plesent wether middle and latter part of the day quite calm saw and lowered for Spirm Whales this afternoon about 2 clock but got none. the whales saw the boats and wher O. P. H. — but the starbord boat got fast with only one iron and that rather slitly and soon drawed and was off. we continued to chase them until dark but did not succeed in getting fast again hard luck and nothing for it but fatigue and back ake so ends this day of our Lord and Savour

Tuesday August the 19 [in the Mozambique Channel] . . . saw and lowered for spirm Whales but to no effect the Whales standing N W the boats did not se them after they lowered

Wenesday the 20 . . . fine wether saw and lowered the boats for spirm Whales and got one got him alongside and commenced cutting about 1 clock got him all in but his head by dark and left that until morning — saw a sail standing S E — got supper and started the works — so ends this lucky day of our Lord Whales standing N W

Thursday Aug 21 . . . all hands employed in boilling and clearing away the head — saw whales this morning about 4 miles of — when we was heaving in the head and as soon as we could get ready we run down for them but did not se them again all hands up to the eyes in fat

Sat August the 23 . . . finished boilling last evevening—have employed today in stowing down the oil

Satterday Aug the 30 . . . saw and lowered the boats for a large Spirm Whale and the wast boat got fast and soon got loose again on account of line getting cut of by the first iron as the first iron was not in the whale but was darted at his head but did not enter. the boats continued to chase him for 3 or four hours but to no effect as he went very fast to windard so much for this unlucky day of our Lord and Master

Satterday Sept the 6 . . . saw and lowered the boats for Spirm Whales and the larbord boat got one. took him alongside and commenced cutting. cut until dark and then left him until morning — shorting sail and got supper and then set the wach — Cut of the Whale 63 feet

Sunday Sept 7 . . . all hands employed in cutting and heaveing in the whale — got the bodey all in at dark and left the head until morning — got supper — set the wach — and started the works — so ends this sabbath day of our Lord

Tuesday Sept the 9 . . . all hands busey and fully employed. saw and lowered the boats for Whales and chased them but to no effect

Wednesday the 10 . . . employed in boilling — lowered the boats for Whales. The Starbord and Larbord boat got one each

Thursday Sept the 11 . . . all hands up to there eyes in fat-boilling and stowing down lowered the boats for Whales. The Larbord and Waste boats got one each the Starbord boat got fast also but lost there Whale on account of one of the irons drawing and the other breaking off — the Waste boat got slightly stove—got them alongside about sundown and cut one of them in that we got yesterday and then got supper. set the wach and commenced boiling. so ends this luckey day.

Friday Sept 12 . . . employed in cutting in and boiling

Sabbath Sept the 14 . . . got through boilling this morning. all hands busey in stowing down the oil

Wedsday Sept 17 . . . saw Whales. lowered the boats and chased them but to no effect as they was a goin very fast and it being very ruged

Friday Sept the 26 . . . saw Whales today and run for them but they did not prove to be the right kind

Sunday [Oct] 5 [bound in port] . . . lower and chased blackfish and that was our share of them

Wednesday Oct 8 . . . saw the land this morning and run for it and come to Anchor at half past 12 at night in Nos Beh

Thursday Oct 9th . . . the boat went on shore early this morning on business to the Governor returned about noon when we whead anchor and run over to Passandava Bay to get our wood arrived there and come to anchor at 7 P M in eight fathoms of watter

Oct the 12 . . . all hands have ben employed the last

2 days in wooding today most of the crew have been on shore part of the time. latter part of the day broak out for watter and chopping wood on bord

Monday Oct 13 . . . all hands employed in wooding

Tuesday Oct 14 . . . employed this day in wheying anchor and running over to Nos Beh arrived there and come to anchor at 3 P M

Sunday the 19 . . . have ben employed the past week in painting ship and getting watter and one wach on shore part of the time

Wednesday Oct 22 . . . all hands this day have ben employed in getting of watter and stowing it away and sending royal yards alof

October 24 . . . all hands employed in weaighing anchor and getting ship under way got under way at 8 A M . . . got the anchors on the bows and chains stowed below . . . saw a french steam ship bound in to the bay

Tuesday Oct 28 [bound on a Cruise] . . . lowered one boat for blackfish but did not get fast

Monday Nov 10 [bound up Channel] — One Year Out today

Friday Nov the 14 . . . Eclipse on the Moon beginning 10 minets after 2 and continuen until after sun rise — difference of time at home and here 7 hours and 40 minutes

Sunday Nov the 16 . . . saw and lowered the boats

for Spirm Whales and chased them but to no effect a sail in sight her boats fast to Whale

Sunday Nov the 23 [south end of Madagascar] . . . saw a Galliteen with her courles [colors] half mast and went on bord of her and suppleyed her Captain with medercene as he was sick

Tuesday Nov 25 . . . saw and lowered boats for Spirm Whales and chased them but to no effect after noon lowered again and met with same success . . . two sail in sight in the evening spoke Bark Hope of New Bedford 6 months out 200 bls with 2 — 70 bls Whales she had taken within 3 last days

Thursday Nov the 27 [off Port Dauphin] . . . saw Spirm Whales about one hour before dark about 4 miles off but did not lower the boats as we did not se them again until sundown . . . the land just in sight

Satterday Nov 29 . . . saw and lowered the boats for Spirm Whales and chased and chased them for hours but alas it was no go as the whales were very wild and shy for they have just come on the ground

Friday Dec 5 [on the cruising ground] . . . saw Spirm Whales and lowered and chased them but it was no go — after noon lowered the boats again. the Starbord and Wast boat got a whale each but did not get them alongside till 12 clock

Satterday Dec 6 . . . employed in cutting in the Whales got yesterday got the bodeys in at dark and left the heads until morning. evening started the works and spirm whales in sight

Sunday Dec 7 . . . employed in boiling and took in the heads

Monday December the 8 . . . all hands employed in boiling and up to there eyes in fat

Wensday the 10 . . . this day employed in stowing down oil got through boiling at daylight this morning

Monday December the 15 — hevy gales from the S E and squaley very bad sea on. at 11 A M a sea struck and carred away our larbord bullworks and done some other trifeling demage. took in the wast boat and turned up the larbord boat — in takein in the wast boat carred away the dayes and barers — at sundown gale broke

Tuesday the 16 . . . employed . . . in repearing bullworks

Thursday the 25 Christmas . . . lowered boats at ½ past 7 this morning for whales starbord boat got fast and got stove but hild on until near night and had to cut from him as the boats was out sight of the ship. the ship lost sight of the boats at ½ past 10 did not see them again until about sun down when they got on bord at 8 o clock

Satterday the 27 . . . spoke Bark Hope New Bedford 400 barrels and Ship Susan of N Bedford in sight and boiling . . .

Thursday January the 1 [1846] . . . 2 sail in sight and spoke one of them wich was the Bark Layfietta of Warren — 5 months out 125 barlles

Sunday the 25 [off Port Dauphin] . . . spoke Bark Hope New Bedford 470 bls

Thursday [February] the 12 [bound to the westward] . . . lowered for blackfish — and got none . . . Spoke the Bark Charleston Packet of New Bedford Capt Howland. 20 monnths out and 750 bar

Satterday [March] the 7 . . . arrive in the [St. Augustine] Bay and comed to Anchor at 6 P M

Sunday March the 8 [in St. Augustine's Bay] — fine wether and as hot as —— and no merstake — riding safely at anchor — doing nothing today but tradeing with the natives — and the decks covered with them — Ship Susan in port of N Bedford

[Monday] 9 . . . all hands employed in breaking out and coopering the oil

Tuesday 10 . . . employed in coopering and finished at 4

Thursday 12 . . . evening squaley and blowing fresh got more chain on deck

Sunday 15 . . . one wach a shore on liberty Sailed this morning Ship Susan on a cruse

Monday 16 . . . got off one raft of watter and washed ship redy to paint

Tuesday 17 . . . got of two boats load of wood and painted larbord side ship

Wednesday 18 . . . got of two boats of wood today

cut it up and scrape ship. very hot wether

Thursday 19 . . . got one load of wood and finished painting ship

Friday 20 . . . very hot employed in stowing down watter and wood and getting a raft to go a shore with in the morning

Satterday 21 . . . got of a raft of watter and stowed it down and stowed down wood

Sunday March the 22 . . . at 8 A M all hands called to maned the windlass got ship under way at half past 9 . . .

Tuesday 31 [off Port Dauphin] . . . Spoke the Bark Hope of New Bedford 450 SPIRM

Thursday [April] 2 . . . Spoke Bark Frances of New Bedford 8 months out and 350 Spirm 110 Whale

Friday April the 3 [bound eastward] . . . 3 sails in sight merchantman stearing W N W

Satterday the 25 [on the ground] . . . 4 mounths this day sence we have seen the Spout of a Spirm Whale

Wednesday the 29 . . . saw Spirm Whales goin quick to the windard but did not lower the boats

May the 1 . . . raised a breech 6 or 7 miles of could not tell what it was — latter part saw another breech and spouts 5 miles of called him a humpback

Sunday May the 3rd . . . saw and lowered the boats for Spirm Whales — Wast boat got fast killed and sunk him — the other boat continued to chased the Whale but did not succeed in getting fast — latter part lowered the boats again but of no avail could not get fast — so much for working on the Sabbath

Sunday the 10 . . . 18 Months from Bristol today

Wensday 13 . . . first part saw two large Spirm Whales lowered the boats and the Starbord boat got fast with only one iron that soon drawed and then the whale was O P H to texex — latter part plesent and fine wether rased a school of Spirm Whales four miles of got within about three miles of them with the ship and so near night that we did not lower the boats

Wensday 20 . . . lowered for Spirm Whales and gallied them and they were O P H — latter part lowered again and the boats did not se him as the Whale came up to windard of the ship and boats was to leeward of the ship

Sunday 31 . . . saw the land [Sandy Island] at 4 P M passed on N W side of it within 1½ miles of it

Friday [June] 12 . . . 19 months out today

Sunday the 14 . . . a sail in sight a Frenchman

Thursday 18 . . . saw the Land [north end of Madagascar] four ponts on the larbord bow and lufed up 2 ponts for to head more for it at 12 M double Cape Amber distent from the land about 11 miles and continued through the day following the Land along. at sun down distent from the Land about 4 leagues

Friday the 19 . . . have ben runing for Nos Beh Island today. run until 7 P M and then tacked ship and let her lay with the main yard aback. distent from Nos Beh Island about 5 miles

Satterday the 20 . . . 2 Leagues from the Land — and sent 2 men on shore that we ship here as there time was up

Tuesday 23 [bound for Majambo Bay] . . . commence standing whole waches

Sunday the 28 — first part plesent wether and gentle gales from of the Land — at 5 A M of the mouth of the [Majambo] Bay. middle part the same Land distent about 5 miles and a strong current setting out of the Bay — latter part of the day light ars at 4 P M about 3 miles up the bay and come to anchor as the current was setting out very strong and cold not make eney headway — anchor in 15 fathoms of watter muddy bottom. at 8 P M ded calm and dropt best bower in 15 fathoms — 7 miles up the Bay clewed up all sail and let them hang and turned in

Monday the 29 — this day commences with fresh gales a blowing down the bay. at 2 A M called all hands and put a reef in each topsail and weghed anchor. at 4 o clock was under way with the wind ded ahead but the current setting up the bay — at 12 M light ars and tide runing out and we dropt mud hook in 5 fathoms about 20 miles up the bay — at 3 P M maned the windlas and took the anchor and made sail and soon after discovered that we was in the wrong shop and consequently we about ship in a jiffey and was O P H out of this at 10 clock went out of the mouth of the bay with a plesent gale from the S E

Tuesday June the 30 . . . come to anchor of Majunga pont. as it was light winds and the current runing at the rate of 3 nots and we cold not make eney headway we come to anchor abrest of the Fort about 1 mile from the shore in 10 fathoms of watter with our small anchor — and had not ben thus situated more than a half hour before the cable parted and we dropt about one mile before get the other Anchor down wich when we did it held her fast and secure — at 2 P M tide about ebb and maned the windlas at half past 2 got the anchor catheaded and made sail and beat in to the harbor and dropt mud hook about half a mile from the shore in 5 fathoms of water and muddy bottom

Wednsday [July] the 1st [at Manjunga] . . . at 8 A M the boat went on shore and learnt that we was not permited to get wood or eneything else — as for watter there was none to get — at 11 o clock maned the windlass and took the anchor and made sail — and as the tide was runing out very strong and getting us right on shore we dropt anchor again and clewed up the topsails — at 4 P M ebb tide and maned the windlass and in 20 minets had the anchor catheaded and all sail on and was under way for Nos Beh

Friday 3d . . . at 7 P M dropt mud hook in and on the S.west side of Passandav Bay in 11 fathoms of watter and muddy bottom and one mile from the shore where we intend to get our wood

July the 4 — the great day of indipendence. this day begins with us — by leaveing our bearths at daylight to prepare ourselves to celebrate this great and glorius day. at 7 got breakfast and then went on shore and finished the day a cutting wood and getting it of to the ship — glorius time and no merstake calm through the day and hot as the———

Monday the 6 . . . one wach on shore cutting wood and the other employed in getting it of to the ship — and skrapeing ship

Tuesday the 7 . . . got through our wood today

Wednesday the 8 . . . took our anchor made . . . sail for French Town on the east side of the Bay where we are going to get our water and cork ship and paint and get fresh recrutes — distent across the Bay 18 miles

Thursday the 9 . . . one wach on shore on liberty

Sunday 12 . . . one wach on shore on liberty and the other employed on . . . bord

Wednesday 15 . . . one wach on liberty and other employed in stowing down watter and wood and corking ship

Thursday 16 . . . one wach on liberty and the other employed in corking and cold-taring the ships bens

Sunday 19 — fine wether and warm part of the crew on liberty

Monday 20 . . . all hands employed in getting a raft of watter and painting ship

Tuesday 21 — employed this day in stowing down watter and wood and doing sundry other jobs

Wednesday the 22 [bound on a cruise] . . . called all hands at 4 A M to weigh Anchor and get the ship under way — got under way at ½ past 6 — also got under way at the same time Bark Hector of Warren — both to cruse in the Mozambique Channel

Thursday 23 . . . two sail in sight — one Hector, Warren

Sunday the 26 . . . saw blackfish but did not lower the boats

Friday the 31 . . . the Land [the African coast] in sight also two sails in close to the Land at Anchor

Wednesday [August] 5 . . . 8 months today sence we have taken a Whale

Monday the 10 . . . 21 MONTHS OUT TODAY WITH 300 SPERM AND 135 OF WHALE OIL

Friday 14 [in Mozambique Channel] . . . lowered the boats for a Sulfer Bottom but did not get fast

Tuesday 18 [off Juan de Nova] . . . lowered the boats for blackfish. got none

Thursday [September] 3 [off Africa] . . . Land [Fogo Island and Fire Island] in sight

Satterday the 5 [bound up the Channel] . . . 9 MONTHS TODAY SENCE WE HAVE TAKEING ENEY OIL

Wednesday the 9 . . . saw Whales earley in the morning. at 9 o clock lowered the boats and chased them all day but did not succeed in getting fast

Thursday the 10 . . . this day 22 months out with 300 bls of Spirm Oil and 135 bbls of Whale Oil

Thursday 24 . . . lowered for Black got none Land in sight commenced standing whole waches

Friday 25 [bound to the south] . . . lowered for Black Fish got none. and the Land in sight

Monday [October] the 5 . . . have taken no oil for the last ten months

Satterday the 10 . . . 23 MONTHS OUT WITH 300 bls of sperm and 135 of whale OIL

Wednesday the 14 [off the Cape of Good Hope] . . . spoke Brig Ann of Liverpool . . . a sail in sight

Monday the 19 [South Atlantic Ocean] . . . saw a Right Whale goin very quick and so did not lower the boats. the whale goin to the south

Monday the 26 . . . saw and lowered the boats for a Spirm Whale and did not get him

Thursday 29 . . . lowered the boats for a right Whale and as he had such long legs the boats could not catch him — he was bound for texex

Thursday [November] the 5 . . . Have not takeing a drop of oil for the last eleven months

Tuesday the 10 [cruising for right whales] . . . Spoke the Ship Waverly of New Bedford . . . Emigrant two years out today with 300 bbls of Sperm oil and 135 bbls of Whale Oil

Thursday the 12 . . . lowered the boats for a Grampus but did not get fast

Sunday the 15 . . . lowered the boats for a Sulpher Bottom but could not get fast to him

Wednesday the 18 . . . saw a English merchant ship stearing S S E

Thursday the 19 . . . 3 ships in sight

Satterday December the 5 . . . nothing in sight. 12 Months today scence we have takeing eney Oil

Monday the 7 . . . saw the Land [St. Helena] at 8 A M

Tuesday the 8 . . . passed on the east side of the Island of St. Helena distent from the land about 1 league. latter part 3 ships in sight — commenced standing whole waches

Wednesday the 9 . . . four ships in sight

Thursday the 10 . . . 3 sails in sight . . . 26 Months out today with 300 bls of Spirm and 135 bls of Whale Oil

Monday the 14 — fresh breezs and good wether and sail in sight — at 2 P M distent from the Island of Ascension 73 miles

Tuesday the 15 . . . saw the land at 2 A M — at 12 M got up the chains at 1 P M kept of the Port distent 2½ miles and a boat borded us from the shore and we lernt that we could not get eney watter there and consequently we did . . . not come to anchor but was O.P.H. . . . stowed the cables below so ends this day of disappointment

Friday the 18 [bound for the Equator] . . . saw a Hump Back but did not lower for him

Wednesday the 23 [on the Equator] . . . 2 sail in sight

Monday the 28 [homeward bound] . . . saw Blackfish lowered the boats and got one

Tuesday [January] the 5 [1847 . . . 13 Months today scence we have takeing eney Oil

Sunday the 10 . . . 26 Months out today with 300 bls of Spirm & 135 bls of Whale

January the 17 — first part of this day plesent breezes from the S W — middle part the wind hawled sudenly to the N and blew very hard accompined with a plenty of rain furled every thing but close reefed maintopsail and foretopmast staysail — latter part hawled to the N E by E and quite modderate — made sail and went on our way rejoyceing

Monday the 18 . . . saw a Brig and a sconer stearing to the S S W. latter part . . . 3 sails in sight run quite handy to a Brig and she showed a bleu signal with a letter K in the center

Friday the 22 [off Cape Hatteras] . . . very hevy hail squall lying under goose wing maintopsail and foretopmast staysail — middle part rather more moderate and plenty of hail squalls but not so hard as the first of the day

Satterday the 23 . . . Spoke the Bark Express of Wiscaset

Monday the 25 [on the coast] . . . at 10 A M supposed to be on the N side of the Gulf as there was a sudden change in the wether and watter . . . under all sail stearing north

Tuesday the 26 . . . Spoke the Bark Carlos of Providence 24 hours out and she stearing S W . . . good wether

Wednesday the 27 . . . a 5 A M the wind hauled to the N N W and blew a gale accompined with rain and snow took in all sail but close reef maintopsail and foresail — ship heading to the N E. middl part blowing quite fresh and give her the foretopsail close reefed. saw a ship standing W. latter part gale increasing and a very bad sea on and frezeing wether took in the foretopsail and reefed the foresail — at 7 P M 35 miles S E Montog point

Thursday the 28 — this day begins with the wind from the N N W and blowing a gale and a hevy sea on and very cold wether — ship heading to the W — the deck and riggen covered with ice. lying to under close reefed maintopsail and reefed foresail — middle part more modderate. latter part wind light and from the NW. made all sail plesent through the day and very cold

Friday the 29 . . . at ½ past 7 o clock saw Block Island one point on the larbord bow. ship heading N N W. at eleven o clock the wind quite light. at twelve took a pilot on bord. at 2 P M the wind light at ½ past 2 the wind hauled to the S E and had quite a breeze and with it rain and some fogey. at 6 off brest fort Adams at half past 6 dropt mud hook in New Port Harbor in 6 fathoms of water with 25 fathoms of chain out — wind E S E and rainy — so ends this luckey day

Satterday the 30 . . . at half past 7 maned the windlass and heved in about 25 fathoms of chain when the wind sudenly freshened and blew a gale and had to give her the chain again. middle part blowing a gale from N W and give her more chain latter part still blowing a gale from N W and give her the remainder of the chain wich is 95 fathoms of chain out. at 9 P M more moderate and cold

Sunday the 31 — this day begins with the wind from the N W and very cold and fresh and the wether clear: and the old Bark rideing safely at anchor with 80 fathoms of chain out. at daylight maned the windlass and heved short. at half past seven sheated home the topsails and hoysted them aloft — at eight clock maned the windlass and broke ground but immediately come to anchor again as the wind freshened and we could not get out. clewed up the topsails again and let them hang — middle part modderate breeze and wind from the same quarter — at four o clock furled the sails. wind N N W and light latter part light winds and calm. so ends sabbath

APPENDIX IV

Excerpts from the Journal of Seth F. Lincoln, Boatsteerer on the Emigrant

Sunday Nov 10th [1844]: Left Bristol R I in the Barque Emigrant Capt James Shearman bound on a whaling voyage to the Indian Ocean fitted for 24 mos At 11 A M run on a rock in Newport harbor At 5 P M got off without damage At ½ past 5 came to an anchor And so ends this day

Monday Nov 11th: First part of these 24 hours light breezes from the S W with thick foggy weather Latter part strong breezes

Tuesday Nov 12th: First part of these 24 hours light breezes from the eastward with cloudy weather At 8 A M took our anchor At 10 A M wind hauled to S E with a strong breeze and continued through the day

Friday Nov 15th: First part of these 24 hours strong breezes from N E Steering E S E Middle and latter part light breezes with fine weather All hands employed with ships duty

Sunday Nov 17th: . . . Latter part fresh gales from S W At 9 P M hove her to under close reefed main-topsail and staysails At 11 P M took in the waist boat slightly stove

Monday Nov 18th: First part of these 24 hours fresh gales from S W At 8 A M the wind hauled to N W At 4 P M it began to moderate and put her before the wind and run about half an hour when we hove her to as it was too rough to run At 8 P M kept her off again and made sail steering E S E

Saturday Nov 23d: First part of these 24 hours light breezes from the westward Middle and latter part fine breezes from S W with fine weather At 3 P M saw a ship steering N W All hands employed with ships duty Steering S E by E

Thursday Nov 28th: First part of these 24 hours light breezes from the southward with fine weather Saw three sails at Sunset steering to the westward Middle and latter part light breezes with squally weather Steering S E when she will go it

Friday Nov 29th: . . . At 11 A M saw a Brig steering to the westward

Thursday Nov 30th: . . . At 10 A M saw Grampuses and nothing more in sight

Thursday Dec 1st: . . . At sunrise two sails in sight steering to the westward At 1 P M saw a ship steering W N W

Monday Dec 2d: . . . At daylight saw three sails steering W N W

Thursday Dec 3d: First part of these 24 hours light breezes from the S S E with fine weather At 1 P M saw a Brig steering N E At 2 P M saw a sail to the westward . . . All hands employed with ships duty

Friday Dec 6th: . . . At 1 A M saw spouts to the windward called them humpbacks At 11 A M saw a sail to the southward At 3 P M saw more spouts

Saturday Dec 7th: . . . At daylight two sails in sight . . . At 11 A M saw a ship steering S W At 1 P M saw a Brig steering W N W At 4 P M saw an English Brig steering N E

Thursday Dec 8th: . . . At 8 A M saw a ship steering S E At 11 A M saw finbacks At 4 P M lowered to try the boats At 5 P M took the N E Trades

Monday Dec 9th: . . . At daylight saw a Brig standing to the southward At 1 P M saw finbacks

Tuesday Dec 10th: . . . At daylight saw a Brig standing to the southward At 11 A M saw an English Brig steering N N W

Wednesday Dec. 11th: . . . At daylight three sails in sight

Thursday Dec 12th: . . . One month out and no whales got

Saturday Dec 14th: . . . At 8 A M saw a brig standing to the northward At 5 P M saw a sail standing to the northward

Tuesday Dec 17th: . . . At 2 P M saw the land the

island of Bravo one of the Cape De Verds At 7 P M tacked ship off shore

Wednesday Dec 18th: . . . At 3 A M tacked ship and stood in for the land At 1 P M Captain went on shore and sent the boat back One ship in sight

Thursday Dec 19th: . . . At 8 A M the boat went on shore trading At 4 P M the boat returned and we took her up

Friday Dec 20th: . . . At 5 A M the boat went on shore At 4 P M the Captain came on board and kept her off

Tuesday Dec 22d: . . . At 5 P M saw Blackfish going quick to the S W

Monday Dec 23d: . . . Nothing in sight

Wednesday Dec 25th: . . . Nothing in sight

Thursday Dec 26th: . . . Nothing in sight

Saturaday Dec 28th: . . . At daylight two Barks in sight

Thursday Dec 31th: . . . At daylight one Bark in sight steering to the southward and one Brig steering N W At 11 A M saw a sail to the southward At 12 M raised Blackfish lowered and the Starboard boat struck one and killed him At ½ past 2 P M came on board and took up the boats

Thursday Jan 2d [1845]: . . . Boiled out our Blackfish and he turned up 15 galls

Friday Jan 3d: . . . At 6 A M saw Finbacks At 11 P M Old Neptune came on board and after the usual allowance of shaving he took his departure

Saturday Jan 4th: . . . At 3 P M saw killers

Sunday Jan 5th: . . . At 5 A M saw a finback At 11 A M saw a sail steering to the N W

Saturday Jan 11th: . . . At 2 P M saw a Brig steering to the northward At 5 P M saw Blackfish and lowered all three boats . . . came on board without success

Sunday Jan 12th: . . . At daylight saw a schooner steering S W At 8 A M saw breaches called them humpbacks At 5 P M saw a ship steering N W

Tuesday Jan 14th: . . . At daylight saw a Brig standing to the eastward At 7 A M saw porpoises

Saturday Jan 18th: . . . At 8 A M saw a Bark steering S W by W At 10 A M saw porpoises

Thursday Jan 19th: . . . Plenty of porpoises in sight

Monday Jan 20th: . . . At 3 P M saw grampuses

Saturday Jan 24th: . . . Saw grampuses and killers

Monday Jan 27th: . . . At 1 P M spoke Brig Edward Southworth of Mattapoiset 3 mos out with 200 blls spm oil

Wednesday Jan 29th: . . . At 3 P M spoke Bark Superior Bishop of Sag Harbor 18 mos out 2150 blls 1500 spm bound home She took a right whale Tuesday

Saturday Feb 1st: . . . At 2 P M saw killers lowered two boats without success At 4 P M saw finbacks

Sunday Feb 2d: . . . Saw plenty of finbacks through the day At 3 P M saw killers At ½ past 3 saw a sun-fish lowered a boat and took him

Monday Feb 3: . . . plenty of porpoises in sight At 10 A M saw a Brazilian Brig steering W by S . . . plenty of finback in sight

Tuesday Feb 4: . . . At 11 A M saw a Bark standing N W At ½ past 1 P M raised right whales At 2 P M lowered all three boats and the Starboard boat struck and killed him At 4 P M commenced cutting

Wednesday Feb 5th: . . . At 1 A M finished cutting At 6 A M started the works . . . Plenty of finbacks in sight

Thursday Feb 6th . . . At sunset cooled down

Friday Feb 7th: . . . At 7 A M started the works At 2 P M finished boiling

Saturday Feb 8th: . . . Stowed down 43 blls of oil

Monday Feb 17th: . . . At 5 A M saw a Bark steering N N E . . . At 9 A M raised a right whale At 10 A M lowered two boats and the waist boat struck him At 12 M lowered the Starboard boats and killed him and at 5 P M took him alongside and commenced cutting

Tuesday Feb 18th: . . . At 2 P M finished cutting and started the works

Wednesday the 19th: . . . All hands employed with cutting blubber and trying out

Thursday the 20th: . . . At 5 A M cooled down and at 1 P M started the works again All hands employed with cutting and boiling blubber At 5 P M saw a Brig standing N W

Friday Feb 21st: . . . At 6 A M finished boiling Stowed down 60 blls of oil

Saturday Feb 22d: . . . Stowed down 40 blls of oil

Tuesday Feb 25th: . . . At 8 A M raised a right whale and lowered the larboard and waist boats without success At 9 A M came on board and took up the boats

Thursday Feb 27th: . . . At 2 P M raised right whales and lowered all three boats without success

Saturday March 1st: . . . At 2 A M saw a sail standing S S E . . . Finbacks killers grampuses and porpoises in sight

Wednesday March 5th: . . . At ½ past 12 P M saw a large sperm whale going quick to the windward

Monday March 14th: . . . At 9 P M saw Blackfish

Tuesday March 18th: . . . At 3 P M saw finbacks

Sunday March 30th: . . . At 5 P M caught a porpoise

Monday March 31st: . . . At 5 A M saw the land the island of Inaccessible At ½ past 5 saw Nightingale At 6 P M [sic] saw Tristan D Acunha N point bearing E distant 20 miles At 8 A M lowered the waist boat and the mate went on shore trading At 11 A M the boat came on board and at 2 P M went on shore again At 6 P M took up the boat

Wednesday April 9th: . . . Cape of Good Hope bearing N distant 218 miles

Monday April 14th: First part of these 24 hours strong gales from the W by S with a very heavy sea Steering E At 8 A M lost the larboard boat with the craft and davies falls &c At ½ past 8 the Starboard boat went also with craft davies falls &c At 6 P M lost a hog overboard probably he went to look after the boats

Sunday April 20th: . . . At 5 P M saw a sail standing to the westward

Tuesday May 6th: First part of these 24 hours strong breezes from S E . . . At 10 A M carried away foretopmast fore and main topgallant masts . . . At 3 P M saw a Bark steering N W All hands employed in clearing the wreck And so ends this day Hard luck

Tuesday May 13th . . . At 8 A M saw the land the Isle of France bearing N W distant 30 miles At 9 kept off for it At sunset hauled off shore

Wednesday May 14th: . . . The land in sight At daylight kept off SW . . . At 2 P M hauled our wind to the eastward The Isle of France proved to be Bourbon our chronometer being 60 miles to the eastward

Thursday May 15th: . . . At daylight Bourbon in sight bearing S W by W distant 11 leagues

Monday May 16th: . . . At daylight made the land the Isle of France bearing S E distant 12 leagues

Saturday May 17th: . . . Port Louis bearing E distant 8 leagues At 10 A M the Steamer took us in tow At 1 P M moored her in the [illegible] and there let her remain at present

Thursday June 5: At 4 P M unmoored and left Port Louis

Friday June 6th: . . . At daylight Isle of France in sight bearing E and Bourbon bearing W S W

Saturday June 17: . . . Both islands in sight

Wednesday June 11th: . . . At 7 A M raised spirm whales going quick to the windward lowered without success. At 2 P M raised them again lowered and the waist boat struck and after running about three cut from him and came on board

Thursday June 12th: . . . At 10 A M raised spirm whales and lowered all three boats At 11 the starboard struck and killed him and at 2 P M took him alongside and commenced cutting

Friday June 13th: At 2 A M finished cutting At 8 A M started the works

Sunday June 15th: At 12 M finished trying

Monday June 16th: Stowed down 25 blls spm oil

June 17th to 19th: Fresh gales from S E with squally weather Nothing seen

Monday June 20th: . . . At 8 A M kept her off N W by N bound to Nos Beh

From June 21st to 25th fresh gales from S with squally weather Lying to under closereefed topsails

Tuesday June 24th: . . . At 1 P M saw the land about 80 miles to the south of Cape Amber At 3 P M hauled our wind off shore

Wednesday June 25th: . . . At daylight Cape Amber bearing S W by S distant 15 miles

Thursday June 26th: . . . Cape St Sebastian bearing S E distant 12 miles

Friday June 27th: . . . At daylight Nos Beh Island bearing South distant 30 miles . . . At 8 PM came to an anchor at Nos Beh

Saturday June 28th: . . . At 11 A M took our anchor and proceeded to get wood and water we came to anchor again at 5 P M

Wednesday July 9th: . . . At 1 P M took the sea breeze At 5 P M spoke Bark Bruce Daggett of Fair Haven 7½ months out clean Bound in

Thursday July 10th: . . . At sunset Nouk How Island bearing S distant 10 miles

Saturday July 12th: . . . At daylight one sail in sight to the southward At 1 P M crossed the Chesterfield shoals

Monday July 14th: . . . At daylight one ship in sight a French sloop of war

Thursday July 17th: . . . At 5 P M made the land bearing N N W distant 10 miles

Monday July 21st: . . . The land in sight

Wednesday July 23d: . . . At 11 A M saw finbacks

Thursday July 24th: . . . At 1 P M saw humpbacks

Wednesday July 30th: . . . At 10 A M saw grampuses lowered two boats without success At 4 P M saw them again and lowered all three boats without success

Friday August 1st: . . . At 11 A M saw grampuses . . . At 2 P M raised Blackfish and lowered three boats without success

Saturday August 2d: . . . At 9 A M saw the island of Juan D Nova bearing N by E distant 3 leagues . . . Saw finbacks killers and porpoises

Monday August 4th: . . . At 2 P M raised sperm whales and lowered all three boats The Starboard boat struck one and drawed from him At sunset came on board and took up the boats

Tuesday August 5th: . . . Nothing to be seen but grampuses

Saturday August 9th: . . . At 7 A M saw Blackfish going quick to the windward . . . Saw porpoises

Monday August 11th: . . . At 9 A M saw breaches called them Jumpers . . . At 3 P M saw the land Juan De Nova bearing N E distant 7 leagues

Wednesday August 13th: . . . At 1 P M saw the land bearing S E by E distant 7 leagues

Friday August 15th: . . . At daylight saw a Brig steering N E At 3 P M saw Blackfish lowered two boats without success

Tuesday August 19th: . . . At 1 P M saw sperm whales and lowered all three boats without success At 4 P M came on board and took up the boats . . . And so ends this day Hard luck

Wednesday August 20th: . . . At 7 A M raised sperm whales and lowered all three boats and the larboard boat struck one and killed him At 11 A M took him alongside and commenced cutting At 7 P M finished the body and commenced boiling

Thursday August 21st: . . . At 7 A M saw a large sperm whale to the leeward At 10 A M took in the head All hands employed in cutting blubber and boiling

Friday August 22d: . . . All hands employed with cutting and boiling blubber At 10 P M finished boiling

Saturday August 23d: . . . Stowed down 42 blls spm oil

Saturday August 30th: . . . At 7 A M raised a sperm whale and lowered all three boats and the waist boat struck him and cut his line off with an iron . . . And so ends this day Hard luck as usual

Saturday September 6th: . . . At 10 A M raised sperm whales and lowered all three boats and the larboard boat struck one At 1 P M took him alongside and commenced cutting

Tuesday September 9th: At 2 P M saw sperm whales and lowered without success The waist boat got on and darted but did not strike

Wednesday September 10th: At 10 A M raised sperm whales and lowered all three boats The starboard and larboard boats struck and at 4 P M took them alongside

Thursday September 11th: At 10 A M raised sperm whales At 1 P M lowered all three boats The starboard boat struck and one iron broke and the other drawed and lost him The larboard and waist boats struck and killed them and at sunset took them alongside Stowed down 53 blls spm oil

Sunday September 14th: Stowed down 41 blls spm oil

Monday September 15th: Stowed down 56 blls spm oil

Wednesday September 17th: . . . At 9 A M raised sperm whales and lowered all three boats without success At 12 M came on board and took up the boats

Friday September 26: . . . At 8 A M saw spouts and at 10 A M saw breaches but did not ascertain what they were

Tuesday September 30th: . . . At 10 A M saw a sail steering N

Wednesday October 1st: . . . One schooner in sight steering N N E

Thursday October 2d: . . . At 5 P M saw the land

Wednesday October 8th: . . . At daylight saw the land bearing S E distant 20 miles At 11 P M came to an anchor at Nos Beh

Thursday October 9th: . . . At 1 P M took our anchor with a fine breeze from W N W At 6 P M dropped our anchor at Passandava Bay for wood

Tuesday October 14th: . . . At 5 P M dropped our anchor at Nos Beh for water and recruits

Friday October 24th: At 8 A M took our anchor and left Nos Beh with a light breeze from S E At 10 A M took the sea breeze from W S W At 5 P M saw a French steamer bound in

Saturday October 25th: . . . At 6 A M turned up the starboard boat for painting under the superintendence of Mr S Palmer second officer At 8 A M saw Blackfish but did not lower as we had plenty of oil on board

Tuesday October 28th: . . . At 6 A M saw Blackfish and lowered one boat without success

Monday November 3d: . . . Saw plenty of grampuses At 6 P M saw the land the Island of Juan De Nova bearing S distant 15 miles

Wednesday November 12th: . . . This day completes the first year of our cruise We have taken seven spirm whales which made us 220 blls of oil and 2 right whales which made us 140 blls making in all 360 blls Great doings indeed Fortune has indeed favored us so far

Sunday November 16th: . . . At 3 P M raised sperm whales At 4 lowered all three boats but could not get on At sunset came on board and took up the boats Saw a Bark to leeward She had her boats down and fast to a whale but they did not get him

Wednesday November 19th: . . . Saw humpbacks At 10 A M saw the land bearing E distant 20 miles At 5 P M saw the breakers on Star reef

Thursday November 20th: . . . At daylight saw the land Cape St Mary bearing N N E distant 10 leagues

Sunday November 23d: . . . At 4 P M spoke a French schooner from Madagascar bound to the Isle of France The Captain came on board to get some medicine as he was very sick

Tuesday November 25th: . . . At 10 A M raised sperm whales At 11 lowered all three boats but could not get on At 1 P M came on board and took up the boats At 3 lowered again but without success At 4 P M came on board and took up the boats At 9 P M spoke Bark Hope Ellis of New Bedford 6 months out with 140 blls spm oil

Thursday November 27th: . . . At daylight made sail

Saw the land . . . distant 8 leagues . . . At 6 A M raised sperm whales going quick to the windward

Saturday November 29th: . . . At 7 A M raised sperm whales At 8 lowered all three boats and gallied the whales with the ship The Starboard boat got on and darted but missed him At 10 A M came on board and took up the boats At 4 P M saw a bark to the northward She had her boats down in pursuit of whales

Friday December 5th: . . . At 9 A M raised sperm whales At 10 A M lowered all three boats The larboard boat got on and darted but did not strike At 12 came on board and took up the boats . . . At 2 P M raised them again and lowered all three boats The Starboard and Waist boats struck and killed them and at 11 P M took them alongside

Saturday December 6th: At 7 A M commenced cutting . . . At 7 P M finished cutting the bodies At 8 started the works

Sunday December 7th: At 8 A M commenced cutting the heads At 3 P M finished cutting

Monday December 8th: . . . At daylight saw the land bearing W N W distant 5 leagues All hands employed with cutting blubber and boiling

Tuesday December 9th: . . . At daylight saw the land bearing N W distant 8 leagues At 1 P M saw Blackfish All hands employed with cutting blubber and boiling

Wednesday December 10th: . . . Stowed down 70 blls spm oil

Thursday December 11th: . . . Stowed down 14 blls of oil

From Friday 12th until Monday 15th: strong gales from S E with thick rainy weather and a heavy sea running Sunday 14th saw a Ship steering SW Monday 15th at 10 A M shipped a sea which stove our bulwarks on the larboard side but did no serious injury At 4 P M the gale abated and the wind hauled to N W No observation

Tuesday December 16th: . . . Repaired the bulwarks rail &c At 6 P M saw a Ship steering E

Saturday December 20th: . . . At 1 P M saw the Land bearing N W

Monday December 22d: . . . At daylight saw the land . . . No observation

Thursday December 25th: . . . At 6 A M raised sperm whales At 8 lowered all three boats and the starboard boat struck but could not get a chance to kill him At 5 P M cut from him and at 7 came on board and took up the boats . . . A fine days work for Christmas

Saturday December 27th: . . . At 5 P M spoke Bark Hope of New Bedford 7 mos out with 400 spm Saw Ship Susan of N/B to windward boiling

Thursday January 1st [1846]: . . . At 5 P M spoke Bark Lafayette Bowen of Warren R I 6 months out 190 spm Saw Bark Hope to leeward

Tuesday January 6th: . . . Saw a Bark standing E S E

Thursday January 8th: . . . At 4 P M saw Bark Lafayette to windward At 5 P M saw the land

Monday January 12th: . . . And so ends this day Fourteen months out this day This never will buy the child a frock without we do better than we have so far

Thursday January 15th: . . . At 2 P M saw the land At 3 P M hauled our wind to the southward At 5 P M saw a Bark standing N E by N

Monday January 19th: . . . At 11 A M saw the land bearing W N W distant 9 leagues

Sunday January 25th: . . . At 2 P M spoke Bark Hope Ellis of N/B 8 mos out 470 spm

Friday January 30th: . . . Saw spouts to windward At 8 A M saw a Bark to windward

Saturday January 31st: . . . At 4 A M saw a Bark to windward

Tuesday February 3d: . . . At 10 A M saw Grampuses

Friday February 6th: . . . At 11 P M caught a cowfish

Tuesday February 10th: . . . At 5 P M saw Blackfish and lowered all three boats without success

Thursday February 12th: . . . At 7 A M saw Blackfish and lowered two boats without success At 11 A M saw a sail standing S W At 12 N saw grampuses At 4 P M spoke Bark Charleston Packet of New Bedford 20 months out with 450 blls 80 wh bound home

Friday February 13th: . . . Saw a Bark to windward

Saturday February 14th: . . . At daylight saw Blackfish and lowered two boats without success . . . At 2 P M saw Blackfish and cowfish and lowered all three boats and the Starboard boat struck a cowfish and took him on board At 4 P M came on board and took up the boats

Sunday March 1st: . . . At 10 A M saw a Brig steering N W

Monday March 2d: . . . At daylight saw a sail steering W S W

Thursday March 5th: . . . At 11 A M saw the land bearing E N E distant 14 leagues

Friday March 6th: . . . Saw the land distant 7 leagues At 8 A M saw a sail standing N E

Saturday March 7th: . . . Saw the land distant 5 leagues At 5 A M saw Sandy Island bearing E by N distant 4 leagues . . . At 6 P M dropped our anchor in 7 fathoms at St Augustine Bay And there let her lie at present And so ends this day In port Ship Susan Manchester of New Bedford 7 months out with 250 spm

Sunday March 22d: . . . At 8 A M took our anchor and left the bay . . . At 4 P M Sandy Island bore S E distant 8 leagues

Monday March 23d: . . . At 11 A M saw the land bearing E by S distant 12 leagues

Tuesday March 24th: . . . At daylight saw the land bearing E S E distant 8 leagues

Wednesday March 25th: . . . Saw the land bearing E distant 4 leagues

Tuesday March 31st: . . . At 2 P M spoke Bark Hope Ellis of N/B 10 mos out with 470 spm

Wednesday April 1st: . . . One sail in sight

Thursday April 2d: . . . At 11 A M saw a Ship steering S W At 1 P M saw a sail to the southward At 3 P M spoke Bark Frances Tabor of New Bedford 8 mos out with 380 spm 120 wh

Friday April 3d: . . . At daylight saw a Ship steering S W At 10 A M saw a Bark steering S W At 11 A M saw a Brig steering S W

Saturday April 4th: . . . At 1 P M saw an English Brig steering W

Sunday April 12th: . . . And so ends this day 17 months out this day We are on the road to fortune

Monday April 13th: . . . At 10 A M saw grampuses

Wednesday April 29th: . . . At 7 A M saw sperm whales going quick to the windward

Friday May 1st: . . . At 1 P M saw breaches and spouts called them humpbacks

Saturday May 2d: . . . At 8 A M saw porpoises and caught three

Sunday May 3d: . . . At 8 A M saw grampuses At 10

A M raised sperm whales At 11 lowered all three boats and the waist boat struck one killed and sunk him At 1 P M came on board and took up the boats

Wednesday May 6th: . . . At 9 A M saw grampuses

Wednesday May 13th: . . . At 7 A M raised sperm whales At 9 lowered all three boats and the Starboard boat struck one and drawed from him At 2 P M came on board and took up the boats At 4 P M saw them again

Wednesday May 20th: . . . At 7 A M raised sperm whales At 9 A M lowered all three boats without success at 2 P M came on board and took up the boats At 5 P M saw them again and lowered all three boats without success At sunset came on board and took up the boats

Thursday May 24th: . . . Saw a Finback At 11 A M saw spouts

Sunday May 31st: . . . At 9 A M saw killers At 1 P M saw the land Sandy Island bearing N E distant 4 leagues

Monday June 1st: . . . At 4 P M lowered the larboard boat and the mate went on shore at Sandy Island At 6 came on board

Tuesday June 9th: . . . At 8 A M saw a Ship steering N E . . . Saw a finback

Sunday June 14th: . . . At 11 A M saw a French Ship standing S

Thursday June 18th: . . . Saw the land bearing S W distant 15 leagues

Friday June 19th: . . . At daylight Nos Beh bore S S W distant 9 leagues

Saturday June 20th: . . . At daylight Nos Beh bore S distant 3 leagues At 11 A M lowered the waist boat and set two men on shore

Sunday June 21st: . . . At daylight made sail The land in sight bearing E distant 10 leagues Saw a Ship standing S W

Monday June 22d: . . . The land in sight

Thursday June 25th: . . . Saw the land bearing S S E distant 11 leagues . . . At sunset the land bore S E distant 3 leagues

Friday June 26th: . . . Saw the land bearing from S W to E S E

Sunday June 28th: . . . At daylight the entrance to Majambo Bay bore S E distant 9 leagues . . . At 8 P M dropped our anchor in 15 fathom on the east side of the passage

Monday June 29th: . . . At 2 A M took our anchor At ½ past 11 A M dropped our anchor in 5 fathom about 2 leagues up the bay At 3 P M took our anchor At 5 P M kept her off N N W having made a mistake in the name of the bay . . . Steering W S W for Majunga point

Tuesday June 30th: . . . At daylight the land bore

from S W to S E distant 3 leagues . . . At 10 A M dropped our anchor in 10 fathom Majunga point bearing S W distant 2 leagues At ½ past 10 parted the chain and lost our larboard anchor At 3 P M got under way with a light breeze from S S W At 5 P M dropped our anchor in 1 fathom abreast the town and may she rest easy for one fortnight

Wednesday July 1st: . . . At 4 P M took our anchor and left Majunga with a moderate breeze from the S W Majunga has a good harbor but as the embargo was on we could obtain no wood and water or recruits . . . Steering N E bound for Passandava Bay

Friday July 3d: . . . At daylight Nos Beh bore E S E distant 6 leagues At 7 P M dropped our anchor in Passandava bay on the Madagascar side in 10½ fathom

Wednesday July 8th: At 8 A M took our anchor with a light breeze from S E . . .At 4 P M dropped our anchor at Nos Beh

Thursday July 9th: Arrived Bark Hector Martin of Warren R I 12 months out with 300 spm

Wednesday July 22d: At 6 A M took our anchor with a light breeze from the E S E In company with Bark Hector

Thursday July 23d: . . . At daylight Nos Beh bore E N E distant 7 leagues The Hector and one Ship in sight

Friday July 24th: . . . At daylight saw the land bearing E distant 14 leagues

Sunday July 26th: . . . At 8 P M saw Blackfish

Wednesday September 9th: . . . At 7 A M raised sperm whales At 10 A M lowered all three boats without success At 6 P M came on board and took up the boats

Thursday September 24th: . . . At daylight saw the land bearing W distant 11 leagues At 9 A M saw Blackfish and lowered all three boats without success At 10 came on board and took up the boats

Tuesday September 29th: . . . Saw finbacks porpoises and squid

Wednesday September 30th: . . . Saw plenty of finbacks

Thursday October 1st: . . . Saw plenty of finbacks

Tuesday October 6th: First part of these 24 hours fresh gales from S E Steering W by S Middle blows heavy in squalls At 3 P M sent down foretopgallant and main royal yards

Wednesday October 7th: First part of these 24 hours fresh gales from S S E . . . At 8 A M housed mizzen topmast

Thursday October 8th: . . . At 6 A M sent in flying jibboom At 11 A M sent down foretopgallantmast

Sunday October 11th: . . . At daylight saw a Bark steering E

Monday October 12th: . . . At 3 P M saw an English Bark steering E N E

Wednesday October 14: . . . At 2 P M spoke Bark Ann of Liverpool At 6 P M saw a Bark steering N E

Monday October 19th: . . . At 2 P M saw a right whale going quick to the southward

Friday October 23d: . . . At 1 P M saw a sail standing to the eastward

Monday October 26th: . . . At 9 A M saw a large sperm whale saw him but two risings At 2 P M saw him again and lowered all three boats but without success At 5 P M came on board and took up the boats

Tuesday October 27th: . . . At daylight saw a ship steering to the eastward

Thursday October 29th: . . . At 11 A M saw a right whale going quick to windward lowered two boats without success At 12 came on board and took up the boats and kept on our course

Tuesday November 10th: At 11 A M spoke Ship Waverley of New Bedford Steering E trying out She saw right whales in the morning

Wednesday November 18th: Fresh gales from S W with thick weather Steering N Saw English Bark Factory steering E N E

Sunday November 22d: Fresh breezes from S E with thick weather Nothing to be seen but humpbacks and finbacks Steering N N E bound I know not where

Sunday November 29th: Northeast Trades and a

fine breeze Steering N N E bound to the north Dull music

Tuesday December 1: Moderate breezes from S E with fine weather but no whales Steering N N E under easy sail

Tuesday December 8th: . . . At daylight St Helena bore W N W distant 6 leagues Commenced standing half watches . . . At sunset the land bore S E by S distant 15 leagues

Wednesday December 9th: . . . At daylight three sails in sight steering N W

Thursday December 10th: . . . Two sails in sight steering N N W . . . Twenty-five months out this day Fortune has indeed favored us so far and if she continues her favors a few more years we shall probably end our days in the almshouse or states prison But let us hope on

Friday December 11th: . . . One sail in sight steering N by W

Saturday December 12th: . . . Nothing in sight

Sunday December 13th . . . One sail in sight

Monday December 14th: . . . One sail in sight steering S W . . . At sunset saw the land the island of Ascension bearing N W distant 17 leagues

Tuesday December 15th: . . . At daylight Ascension bore N W distant 4 leagues At 10 A M hauled up for the anchorage with the intention of coming to an anchor At 1 P M kept off N N W as we could obtain no water . . . At sunset the land bore S by E distant 12 leagues

Sunday December 20th: . . . Saw cowfish and porpoises

Monday December 21: . . . Saw a Brig standing S W Saw Blackfish cowfish and porpoises going quick to windward

Tuesday December 22d: . . . Saw humpbacks
Wednesday December 23d: . . .Plenty of porpoises and dolphin in sight

Thursday December 24th: . . . six sails in sight standing S S W

Friday December 25th: . . . Nothing in sight At 3 P M we took the N E Trades

Monday December 28th: . . . At 4 A M saw a sail steering S S W At 4 P M saw blackfish and lowered two boats The waist boat struck one and saved him At 6 P M took him in and took up the boats kept her off her course

Wednesday December 30th: . . . Saw a Brig standing N

Thursday December 31st: . . . Tryed out two blls blackfish oil

Thursday January 7th [1847]: First part of these 24 hours strong breezes from the E N E with squally weather Steering N W At 2 P M took in main topgallantsail for the first time since passing St Helena

Monday January 11th [last entry]: First part of these 24 hours fresh breezes from the E N E with squally weather Steering N W by W Middle and latter part light breezes from S E And so ends this day

Principal Sources

MANUSCRIPT SOURCES

Bristol-Warren Custom House. Papers and documents (in Rhode Island Historical Society).

Ely, Ben Ezra Stiles. Autobiography (property of Ben Ely of Hannibal, Missouri).

Lincoln, Seth F. Journal of Bark *Emigrant* (in Rhode Island Historical Society).

Tucker, Charles F., Journal of Bark *Emigrant* (in Marine Museum at Fall River).

PRINTED SOURCES

Federal Writers' Project (Massachusetts), Works Progress Administration. *Whaling Masters* (New Bedford: Old Dartmouth Historical Society, 1938).

Federal Writers' Project (Rhode Island), Works Progress Administration. *Rhode Island, A Guide to the Smallest State* (Boston: Houghton Mifflin, 1937).

Haley, John William. "The Voyage of the Bark 'Emigrant.' " *"Old Stone Bank" History of Rhode Island* (Providence: Providence Institution for Saving, 1939). III, 203–205.

Hegarty, Reginald B. *Returns of Whaling Vessels Sailing from American Ports* (New Bedford: Old Dartmouth Historical Society, 1939).

Horsburgh, James. *The India Directory, or Directions for Sailing to and from the East Indies, China, Australia, and Adjacent Parts of Africa and South America* (5th ed., London, 1841).

National Archives Project, Works Projects Administration. *Ship Registers of New Bedford, Massachusetts* (Boston Mass., 1940).

New Bedford *Mercury*. October 13, 1849.

Sherman, Stuart C. *The Voice of the Whaleman* (Providence, 1965).

Starbuck, Alexander. *History of the American Whale Fishery* (New York, 1864).

Survey of Federal Archives, Works Projects Administration. *Ship Registers and Enrollments . . . out of Bristol-Warren, Rhode Island* (Providence, 1941).

Ward, R. Gerard. *American Activities in the Central Pacific 1790–1870* (Ridgewood, N.J., 1967).

Whaleman's Shipping List and Merchant's Transcript, 1844–1849.

This third volume in the American Maritime Library

"There She Blows"

has been composed in Monotype Century Expanded by Eastern Typesetting Company and printed by offset lithography by Halliday Lithograph Corporation. The binding is by the Chas. H. Bohn Company.

Published for The Marine Historical Association, Incorporated, by Wesleyan University Press.